No. 2936
$14.95

Science Fair
Developing a Successful and Fun Project

Maxine Haren Iritz
Photographs by A. Frank Iritz

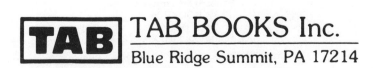

TAB BOOKS Inc.
Blue Ridge Summit, PA 17214

To Michael and Stuart, whose projects
inspired this book,
and
To all Science Fair participants
past, present, and future
Everywhere

*507
IR*

Library of Congress Cataloging in Publication Data

Iritz, Maxine Haren.
Science fair.

Bibliography: p.
Includes index.
1. Science—Exhibitions.   I. Title.
Q105.A1I75   1987   507'.8   87-10030
ISBN 0-8306-0936-9
ISBN 0-8306-2936-X (pbk.)

Questions regarding the content of this book
should be addressed to:

Reader Inquiry Branch
Editorial Department
TAB BOOKS Inc.
P.O. Box 40
Blue Ridge Summit, PA 17214

Cover photographs by Michael Bruce Iritz

2101

Contents

Introduction

As you know, there is currently a great emphasis on raising our educational standards. Because our society is increasingly reliant on technology, the particular stress has been on math, science and computers. Therefore, school districts throughout the country are concentrating energy and resources in these fields.

The showplace for many of these efforts is the annual science and engineering fair, for which students prepare and exhibit science projects. These projects require students to conduct research and experimentation needed to prove a hypothesis. Once completed, these projects, together with displays, are entered in individual school fairs where winners from each category, or scientific field, advance to a regional science fair. The most outstanding projects proceed to state or international competitions, where recognition, as well as scholarships, trips and other prizes, can be considerable.

You're probably reading this book because you (or perhaps your son or daughter) are planning to do a science project, and are looking forward to competing in a science and engineering fair. Although educationally valuable and often emotionally satisfying as well, a science project requires a major commitment of time and effort. Along the way, students usually need not only encouragement and moral support, but some very concrete assistance, too.

Preparation can often take up to a year of intense effort, from the crucial first step, the selection of a topic appropriate to the interests and abilities of the student, to the construction of the display at the project's conclusion. Errors can be costly, both in terms of money spent for supplies and equipment, and the time and energy diverted from the ultimate goal.

Parents are most anxious to assist. They are, in fact, a crucial factor in the success of a project, especially for a first time participant. Unfortunately, many parents, who themselves never did a project and whose scientific background may be limited, often do not know how to help their youngsters. Despite considerable guidance, support and dedication from teachers and advisors, there are many facets of science fair participation which, to the first-timer, are formidable. A teacher may sometimes supervise over thirty projects a semester, and therefore cannot always deal with the day-to-day problems of conducting each student's experiment.

This book will present a clear guide on how to complete a successful science project. Our objective is to lead students and their parents through the steps involved, giving some helpful hints along the way. Teachers, judges, participants, and their families have all contributed their feelings, observations, and best advice. This is intended to guide you down the easiest road to success, and help you avoid the pitfalls.

The appendices were designed to help your son or daughter with that important first step, selecting a topic. In Appendix B, a sampling of projects, we have not always included the top prize winners, but instead, projects from several disciplines, as well as varying degrees of complexity. In this manner, we have attempted to reach students' individual interests, and of, their particular degree of scientific knowledge. We've also included a glossary, summarizing the important terminology introduced throughout the book.

In speaking to science fair winners, the most successful participants had a great deal of help and support from their families. We'd like to help you and your youngster achieve not only success, but enjoyment and satisfaction from the entire science fair experience.

We wish to thank everyone whose help and expertise made this book possible. To those who managed the 37th International Science and Engineering Fair in Fort Worth, Texas, thank you for making our trip productive. To the members of The Greater San Diego Science and Engineering Fair Management committee and Student Advisory Board, most especially Philip D. Gay, Ronald E. Domb, Mary Domb Mikkelson, and Howard L. Weisbrod, and to San Diego science teachers Randy Borden of Bell Junior High School and Gary Oden of Morse High School, we owe a special thank you.

Finally, to all students whose projects are referenced or photographed here, muchas gracias, danke, merci, "dom arrigato," grazie, for helping to make this book a reality.

Thank you all.

Chapter 1

So You're Going
to Do A Science Project!

Right now, your overwhelming feeling is probably panic. A science project seems like a huge undertaking, with many things to be done. You may be thinking that it will take so much time and work, you may never be able to goof off again! Therefore, the question uppermost in your mind may be *Why Do A Project?*

For some of you, the answer might be simply that it's required. If that's the case, the completed project will be a large portion of your science grade. Others might be doing the project because they are interested in some area of science and already have chosen their project idea. Another reason might be the sheer joy of competing in a science fair, matching your abilities against your peers.

For whatever reason you're beginning a science project, this is very likely your very first independent study. Working on your own is one of the most important benefits of doing a project. Although you'll have a great deal of help from parents, teachers and other resources, this project will be entirely your responsibility. You'll control the planning and scheduling as well as executing the research and experiment. Although you will use many of the procedures and techniques that you've learned in science lab, you'll have the freedom to set your own pace.

In the first phase of the project, you will select the subject for your experiment. This process gives you the opportunity to explore the general topics included in the various scientific disciplines. Perhaps, you'll learn about subjects you never thought about before.

Another crucial element is the background research paper. Throughout high school and college, you'll be judged on the ability to write a well researched and presented paper. In fact, for many of you, the research paper will be graded not only by your science teachers, but by your English teacher.

Completing a project will give each participant real "on the job training" in carrying out an experiment according to sound scientific methods. In your school labs, your teachers generally specify the steps to follow for each experiment and provide the necessary materials to work with. They are always on hand to insure that the procedure is carried out safely and properly. When completing your project, *you* must design your experiment, find the required materials, specify the procedures, safely carry out the experiment, and record and measure your observations, making sure that you properly adhere to scientific methods.

When the experiment is finished, you will need to compile and tabulate your results. To do this, you may need mathematical or statistical skills, as well as the ability to graphically show your results. Finally, you'll formulate your conclusions, which will show whether or not your experiment "worked." This will demonstrate your ability to correlate the facts you have researched with the results of your experiment.

In your concluding statement, you can also speculate on the possible benefits to society that your results might present. Here, you can also mention any plans you might have to continue working in that area. Sometimes, students become so fascinated with their subject area that year after year, they continue expanding on their research topic, until, by the time they're 12th graders, they've become experts in their field.

Doing a science project may satisfy the artist in you, too. Creating a display, which will illustrate and present a summary of your project, should show you and your work

Fig. 1-1. Illustration of computer symbol.

to its best advantage. Producing an effective backboard will call on your design, artistic, and photographic skills.

If you've never competed in a science fair before, you might wonder what judges look for. From the school fair to the international level, judges agree that the steps and methods used in the experiment are usually more important than your results or whether you prove your theory.

For those of you who advance beyond your school fair, you'll probably get to spend a few days at the city, county, or state fair. There, guided tours to areas of interest are usually arranged. Professionals are sometimes available for counseling about scientific careers. Finally, you'll also get the chance to meet new and interesting people. In the process, you'll discover, within yourself, a poise and grace you never knew existed.

Although it is by no means necessary to have a home computer to do a successful science project, I have included *computer notes* throughout this book, shown with the symbol in Fig. 1-1. This will suggest the various ways to effectively use your hardware and software. For those of you who do not have your own system, several students said they were able to use their school computer labs after hours or on Saturday.

This book was written for you, the student, to make the science project experience the most enjoyable and educational experience possible. You may doubt your scientific abilities. This is only natural, since this is probably your first undertaking of this type.

Many other students, just like you, have started projects with the same mixture of uncertainty and fear. However, with a lively curiosity and a good measure of diligence they have created effective, scientifically acceptable, award winning projects.

For many of you, the learning will extend far beyond the areas of science to writing, art, mathematics and research. Along the way, I hope you will also find a true sense of pride and accomplishment, and learn not only about science, but about yourself.

Like any other long term effort, a science project is filled with many small victories, false starts, bursts of activity, periods of procrastination, sleepless nights and stomachs full

of butterflies. Through it all, regardless of awards or grades, every youngster who completes a project is a winner.

Remember that the main objective is learning. An exhibitor in the 1986 International Science and Engineering Fair, Stephen M. Jacobs, from Hamilton, Ontario, Canada designed and built a *Continuous/Pulsed Ion Beam Accelerator* (see Fig. 1-2). In the more than two years that it took to build and test the device, he "learned about the atom in every detail and from every angle. Everything that I learned with this project, every application I have identified and made known to others has given me much understanding of the atom and its infinite uses for man."

When Mr. Jacobs wrote about his project, he included a note to all young scientists. "Spending several years on a large scale science fair project may seem rather difficult to comprehend, but if you have a question that you would like answered (be it biological, behavioral, chemical, physical, etc.) the time you spend, answering it for yourself means that you have gained a greater amount of knowledge and experience—more so than if you were to refer to a text. Science has only one purpose—the full understanding of

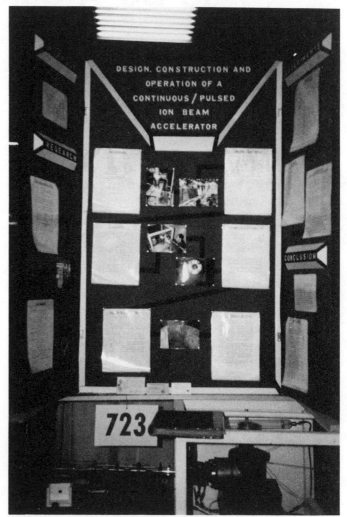

Fig. 1-2. Stephen M. Jacobs; *Design, Construction, and Operation of a Continuous/Pulsed Ion Beam Accelerator.*

everything around us and within us. The projects you do in the future may well have some direct application for humanity. Perhaps you will find a cure for cancer, or maybe you could design some new form of space travel. Whatever you do, complete the project with the full realization that the time you have spent was spent well, and with every further question you ask, you are developing a greater understanding of who you are, where you are and what your purpose is here." Look at your science project not as an assignment but as a journey of discovery. Happy exploration!

Chapter 2

Choosing a Topic

If there is one factor that is critical to the success of all science projects, it is the choice of topic. In fact, some of the exhibitors, especially those competing for the first time, said that finding a subject was the most difficult part of the entire experience. Most students strove for an idea that was original, or at least, not overdone. Everyone wanted a concept that excited their curiosity, but they also wanted an experiment which, they hoped, would not be too difficult.

When looking for an idea, the most important consideration is finding a topic that is interesting and fun for you. Unlike other assignments or research papers you may have, a science project will take at least four months. You will need to do intensive research to gather background information and careful experimentation, in order to scientifically prove or disprove the theory that is the basis of the study. The documentation, analysis and presentation of your results and conclusions will also take considerable time. The work will be easier and far more fun if you're working on a subject you enjoy.

CATEGORIES

Many categories of projects compete in science fairs. They cover all scientific and engineering disciplines, ranging from botany to zoology, and everything in between. If you already have an idea of what your experiment will entail, it would be a good idea to see what category it falls into. If, on the other hand, you have no idea at all, examining the various categories might at least give you some insight into the general area you wish to work in.

A program from a past science fair contains a great deal of valuable information about the different categories. Usually, the projects competing in each category are listed, along with the name and school of each exhibitor. One of the first things you may notice is that some categories have many more entries than others. This, especially at the junior division level, occurs for several reasons. One important reason is that these popular fields are covered in the junior high school science classes. Availability of background information in that particular category is another reason for having more projects in that area. Finally, in certain areas, the subject matter is more relevant to the students' lives, for example human behavior, computers, or medical science. However, these reasons, which make a certain area of study easier and more popular, will also mean stiffer competition in those categories.

Take special notice of concepts that have been presented very frequently. Your topic does not necessarily have to be original for your project to be a success, but teachers and judges will look for a creative approach. Therefore, if you plan to use an idea that has been done often, look at it from a fresh angle. Although information is more readily available in these areas, you may need to work harder to manage an original viewpoint.

Categories, as defined by the Greater San Diego Science and Engineering Fair, and some examples of projects in each grouping are:

☐ Animal Behavior
  —Interaction vs. Back-Up: Testing Two Theories of Mouse Navigation
  —Pitch Discrimination in Goldfish
☐ Biology/Microbiology
  —Bacterial-Fungal Synergism
  —Penicillin's Effect on Reproductive Rate of Volvox
☐ Botany
  —Effects of Ammonium Chloride and Potassium Nitrate on Spirulina

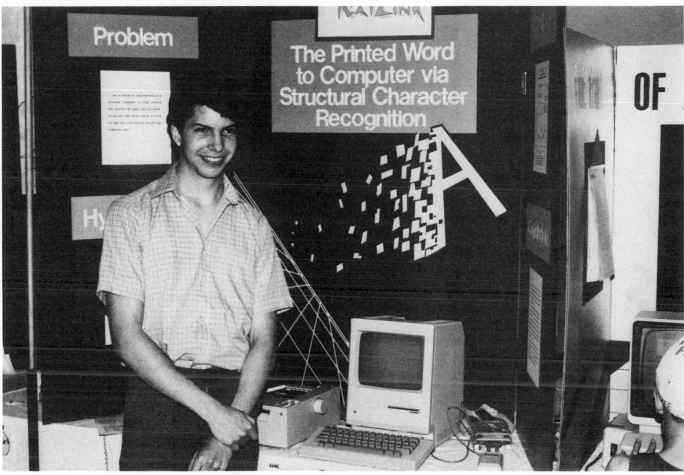

Fig. 2-1. David Calabrese; *TextLink*.

—Hydroponic vs. Soil Gardening
☐ Chemistry/Biochemistry
  Biochemical Cycling by Microbial Life: The Nitrogen Cycle
  —Protein Electrophoresis Determination of Necessary Agents for Planaria Regeneration
☐ Computers
  —TextLink: The Printed Word to Computer via Structural Character Recognition.
  —Yosemite Island: A Computer Simulation of Evolution
☐ Earth/Space Science
  —Shape Analysis of Microfossils Using Laser Shadow
  —Effects of Sunspots and Time of Day on Radio Transmissions
☐ Engineering/Electronics
  —HOT SUB: Drag Reduction by Use of Heated Boundary Layer
  —Which Airfoil Design Will Produce the Most Lift?
☐ Human Psychology/Social Science
  —Racial Stereotypes: What do 2nd and 3rd Graders Believe?
  —Correlation Between Left/Right Brain Hemisphere Dominance and Inkblot Responses

☐ Mathematics
  —Analysis of a Championship Season—the 1984 Padres
  —Randomness in the Game of Life
☐ Medical Science
  —Stress: The Key Element in the Development of Diabetes
  —Self Relaxation vs. REM Sleep to Lower Blood Pressure
☐ Physics
  —Influence of Textured Surfaces on Magnus Effect Rotors
  —Looping Roller Coasters—Can They Be Safe?
☐ Zoology
  —Effects of Shaving on Tarantula Thermotropism
  —Do Pesticides Affect Earthworms?

Categories, as defined by the International Science and Engineering Fair, are somewhat broader. A list of these, together with their descriptions and interpretations, appears in Appendix A.

FINDING A TOPIC

Besides finding a topic that you like, be sure to select

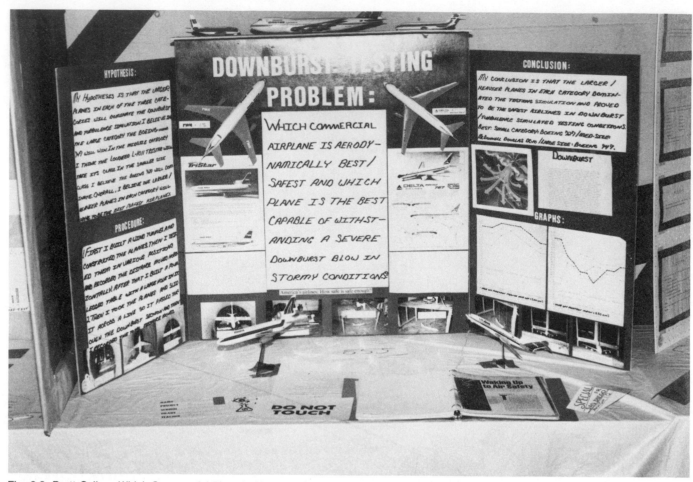

Fig. 2-2. Brett Callan; *Which Commercial Plane is Most Stable in Downburst/Turbulence Conditions?*

a subject with which you feel comfortable and confident. Although the main purpose of doing a science project is to explore new techniques and learn new things, it is best to pick a topic in which you have at least a basic understanding. This is not a good time to do a crash self-study course in your most difficult subject. For example, if mathematics are giving you trouble, a topic requiring complex and sophisticated calculations is unwise.

If this is your first project, or you have not yet taken many science courses, you may be wondering how to find a good idea.

Of course, your math, science, or computer classes are excellent sources. Even classes in social science might suggest an excellent topic idea. For example, a class discussion on subliminal advertising led to an Animal Behavior experiment that measured the effect of sound on mouse appetite. History courses that deal with prehistory might lead to a project dealing with fossils, or dating ancient artifacts.

You might also find a question that arouses your curiosity in a newspaper or magazine. Areas of science, especially those that are important in your particular locality, are often discussed in feature articles. Weekly news magazines deal with the areas of health, aerospace, the environment, and other scientific fields, particularly when new problems, discoveries, or theories are presented.

Problems affecting teenagers are often the inspiration behind projects. The attention on drug-related problems no doubt inspired Doug S. Fay's project, *Drugs Weave Patterns,* which indicates the degree to which alcohol, caffeine, and nicotine impair the spider's ability to weave its web. Catherine Jacobus' project, *The Effects of Selected Carbohydrates on Alcohol Absorption and Behavioral*

Fig. 2-3. Illustration of light bulb.

Modification in Humans, which attempted to find a way to block alcohol's effects, also relates to a very timely issue.

The problem of world hunger inspired Neil Ireland's experiment, which measured the effect of seed size on wheat. Projects on acid rain and pollution levels are also popular choices for many students, for example Leslie S. Thomas' project *Acid Rain—Causes, Effects and Cures, A Seven Year Study,* in which he rehabilitated two ponds that had been previously acidified, or Eric Felt's project, which tested how safe the EPA's standard acceptable levels of toxic wastes and herbicides really were.

Movies or television shows, even if they are fiction, may deal with scientific subjects. For example, several years ago, a successful psychology project used a survey concerning children's reactions to the television film, *The Day After.* More frequent, (perhaps almost to the point of being overdone) are projects dealing with the effects of Rock music (with or without suggestive lyrics), MTV, or video games on heartrate, grades, or a variety of other measurable factors.

A problem, such as a shortage, a natural disaster, an unusual weather condition, or a newly discovered pollutant or health problem, might spark your interest. For example, the energy shortage several years ago inspired J. J. Tessier's project comparing different types of building insulation.

Sometimes, even if you cannot research the condition itself, you might build a project around people's reactions to it, or the psychological impact of a certain issue.

Instances of current, local problems inspiring a science project are the study of the effects of the Tijuana River pollution on the San Diego county beaches, which has become a political as well as an environmental issue in that part of the country. Susan Strickland's experiment, inspired by a local environmental problem rated the flame retardance of various plants that could be grown in the Southern California canyons.

Not only local weather conditions, but the local geography, weather and wildlife itself is often the inspiration for science projects. Living near the ocean usually stimulates interest in the sea itself, as in the Kerry Kostman's project *The Effect of Weather Patterns on Sea Water,* or Ted Larson's marine biology study, *A Comparison of Ocean and Bay Plankton.*

So, the best way to find a topic is to be aware and receptive to your environment. The world around you offers many opportunities for topics.

In talking with science fair participants, we found a variety of reasons for choosing a particular topic.

Fig. 2-4. J.J. Tessier; *Effectiveness of Rigid vs. Batt Insulation.*

Fig. 2-5. Susan Strickland; *Fire Danger from Native Shrubs.*

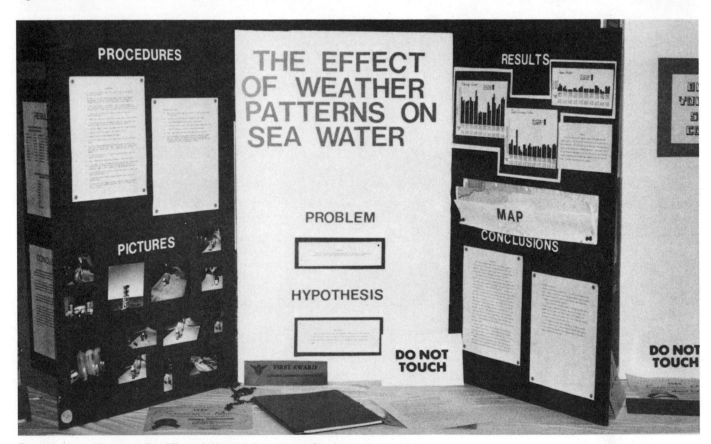

Fig. 2-6. Kerry Kostman; *The Effect of Weather Patterns on Sea Water Composition.*

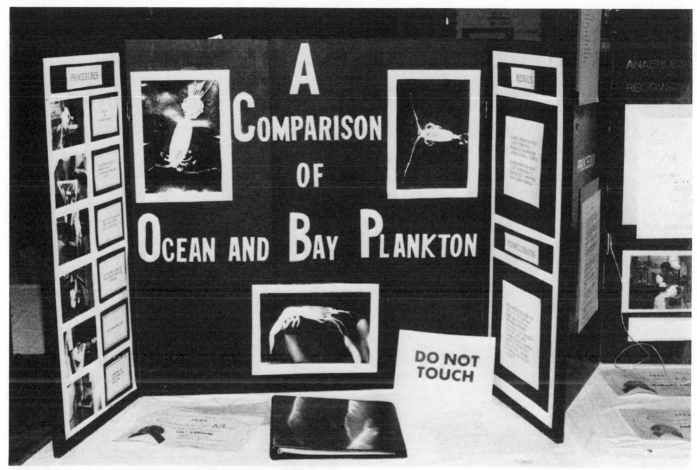

Fig. 2-7. Ted Larson; *A Comparison of Ocean and Bay Plankton.*

"I wanted a relatively simple project with interpretable and quick results," said Nicola Kean of Redlands, CA, discussing her project *What Happens When the Length of a Pendulum is Increased?*

Sometimes, an idea is suggested by observing a particular phenomenon. Maryanne Large of Sydney, Australia, had been unsuccessful in finding a topic when she accidentally observed an effect while printing some photographs at home. From that, she developed her project topic, *Visual Reversal of Photographic Negatives.*

Other concepts resulted from a long standing interest in a particular field. Carl T. Donath of Lafayette, New York, was interested in the idea of a voice or speech recognition system for a computer, and consequently developed a system for *Computerized Speech Analysis and Recognition.* In attempting a project of such scope, Carl had the help of his father, an electrical engineer.

Keith Maggert of San Diego, California is another participant who was inspired by his father's work. When the government denied his father a grant to conduct research to determine the importance of hair to a tarantula, Keith decided to find out. For two years he conducted experiments dealing with the body heat of tarantulas.

Another reason for a choice of topic is to fulfill some particular objective. Ben Cheng, with his project *The Role of Amoebocytes in Zinc Uptake and Transport in the Snail Helix Asperia* wanted a project that would contain factors important to the training of a modern experimental biologist. Although the project was included in the Zoology category, it combined both cellular biology and environmental science.

Some of the participants were quite frank and open about their reasons for choosing a particular subject. "Notoriety and recognition" Billy H. Scott of Jacksonville, Florida told us, when asked why he selected his project *Obtaining Kinetic Energy through Heat,* using rubber bands to 'build' his machine. "I wanted something outstanding, which would attract attention."

A word of warning. Well meaning friends and relatives may propose or try to impose ideas, and even offer their assistance. Before you make a decision, however, *be sure* that *you* are interested in the topic, and have the knowledge and skills to carry it through to completion *on time.* A project that relies solely on the knowledge and work of parents, friends, and relatives will not help you to achieve any of your goals in doing a project. If you are uncertain, discuss the idea with your teacher or advisor. He or she will be glad to talk with you and may even point out something (either

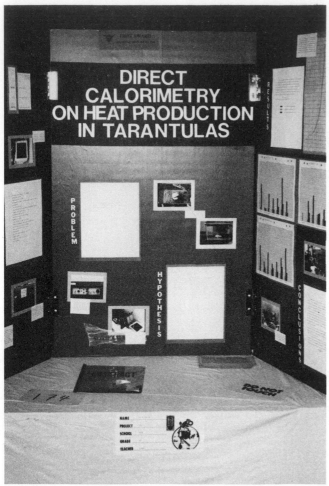

Fig. 2-8. Keith Maggert; *Direct Calorimetry on Heat Production in Tarantulas.*

positive or negative) that you hadn't yet thought of.

CAN I DO IT?

Once you find a topic (or have several possibilities) it's a good idea to analyze the feasibility of your plans. By this, I mean that you must now decide whether you can actually complete the project in the time available. Here's a checklist to help you decide:

☐ Is the information that I need readily available to me?
☐ Where can I find the information?
☐ If the information is not available locally, where is it? How long will it take to get?
☐ Will I need to pay for my information? (government agencies or industry may charge for their data).
☐ If I need special books, can I check them out of the library or must I use them there?
☐ Will I need professional advice? From whom? (specifically!!)
☐ Are they willing to help? Will it cost anything?
☐ What supplies will I need?
☐ Can I build some of the things I need? Do I need help?

☐ Can I borrow some of the supplies and equipment I'll need?
☐ What will my supplies cost?
☐ Do I have the money for my supplies and equipment?
☐ Will my parents let me leave home with their Gold Card?
☐ Can I finish in the time allowed? If not, can I break the project idea into smaller segments?
☐ Is there anything about the experiment my family might object to?

Answering these basic questions will, at least, give you a rough idea of whether you can complete this project. I repeat (at the risk of being boring), if you have any doubts, consult your teacher or advisor. However, if you are sure that this is the topic for you, now is the time to consider some of the elements of your experiment in more detail.

SAFETY FIRST

If you are dealing with chemicals or electricity, you must insure your personal safety as well as the safety of your environment. Some questions that should be answered are:
Electricity:
☐ Is battery power adequate?

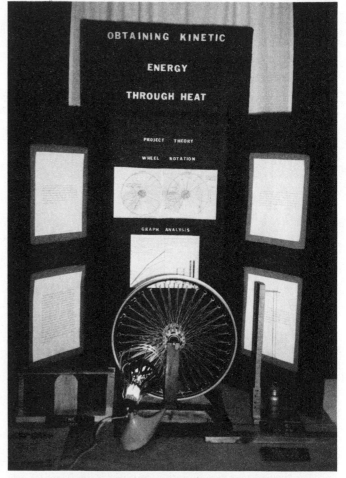

Fig. 2-9. Billy Scott; *Obtaining Kinetic Energy Through Heat.*

☐ If I plan to use house current, is there sufficient wiring?

☐ Can I avoid overloading circuits?

☐ Will everything be properly insulated and grounded?

☐ If I must use line voltage, do I need an electrician?

☐ Do I need the approval of an electrical inspector?

☐ Do I need to notify the local power company?

☐ Is the power the right kind (110 vs. 220 vs. 440) or do I need transformers or converters? Are they readily available?

☐ Is there any danger that I could blackout seven states and the island of Guam?

Chemicals:

☐ Are the substances readily available, or do I need special permits to acquire them?

☐ Is anything I plan to use or synthesize considered a toxic substance?

☐ Do I need approval of state, local or federal authorities (such as the EPA) to do my experiment?

☐ Are there any unstable compounds which could possibly cause an explosion? Will these compounds, either alone or in combination with other substances form toxic gasses, or start fires that might cause injury or property damage?

☐ Will I need special containers?

☐ Do I need protective gear (goggles, aprons, gloves) to work with this substance?

Very possibly, you may need adult assistance or supervision to insure a safe experiment and environment with an electrical or chemical project.

LIVING LABORATORY

Many experiments in the animal behavior and zoology categories rely on the use of live specimens. Although these types of topics may provide some of the most fascinating and fun projects (imagine the thrill of teaching a gorilla to talk!), this introduces another factor into your science project plan.

Carefully consider whether you can and will commit the time and effort to care for these animals. Fortunately for you and your family, the project probably won't involve anything as large or exotic as an ape, but even the smallest live vertebrate requires special handling. Another factor in deciding whether you want to do an experiment of this type is to determine exactly how many specimens you need. You may need a rather large number, kept in separate groups,

Fig. 2-10. Tasha Taylor; *Which Sugar Substances Do Ants Prefer?*

Fig. 2-11. Billy Scott; *Obtaining Kinetic Energy Through Heat* (close-up).

for your experiment to be scientifically valid. Be sure you have the facilities to keep three cages of twenty mice each completely segregated, before beginning this type of experiment.

To do this properly, learn all you can about the animals' care, feeding, habits, and temperature needs, as well as their cleanliness requirements. You must also research the reproductive cycles and characteristics of the species in order to plan for possible periods of gestation and birth during the course of the experiment. For example, when dealing with mice or other rodents, it is extremely important to remove the males when a litter is born, to prevent them from eating the newborns! Many students using rodents have therefore limited their specimens to one sex (provided, of course, that reproduction is not a factor to be considered).

It is also important to understand that some types of laboratory animals are cannibalistic, particularly when they are kept in close quarters. Be sure you can deal with this fact of animal life. If you find it upsetting, choose another topic.

Another word to the wise--be sure to get your parents' approval before embarking on such an experiment. Nothing will terminate your project faster than an unknowing parent entering the house to find 600 (or even 60!) mice running around. Talk to Mom and Dad. You will need their cooperation. If you do get their approval, be conscientious

in keeping the animals and their quarters clean. Help and cooperation will fast disappear in a home that smells like a neglected cage in a travelling zoo!

When working with live vertebrate specimens, be aware of the role of animal rights lobbies. To insure that animals are not mistreated, you must have an expert certify that you are capable of caring for the animals and that your experiment will not harm them in any way. Judges may want explanations of any deaths in your experimental or control groups. Irregularities can prevent you from exhibiting at a science fair.

We have also found, in talking with exhibitors and coordinators from all over the country, that some local fairs use different standards for judging experiments using live specimens. To be sure that your project will be acceptable if you move ahead in competition, adhere to the standards required by Science Service. They conduct the International Science and Engineering Fair. By doing this, you will be able to compete anywhere. You can get a copy of the ISEF Rules by sending $ .50 to:

Science Service
1718 N Street N.W.
Washington, DC 20036

If it turns out that caring for live vertebrates is not for

Chill Out and READ!

you, this doesn't exclude you from working in certain other categories. Try insects. Tasha Taylor won a First Place with an Animal Behavior project entitled *Which Sugar Substance do Ants Prefer?*

On the other hand, you can use vertebrates in a project, while not needing to experiment on them for long periods of time. In *The Canine Allergy-Obesity Connection* by Missi J. Wildenfield, most of the work consisted of recording the dogs' body weight and tabulating these figures together with criteria established for allergic conditions. The only "live" tests on the dogs were pinch tests to determine the body fat content.

DOLLARS AND CENTS

Before making a final decision, estimate all your total costs. It is not necessary to spend a huge amount of money to be successful, nor must you have access to a sophisticated million dollar computer to do a thorough and creative experiment. Some very successful projects, which have advanced all the way to international competition, have been done with minimum expenditure, whereas others require a great deal of money. Students reported that their projects cost anywhere from $12.00 up to $400. Billy Scott wound up spending $150, mostly for light bulbs.

Often, laboratory equipment, as well as computers and other machines, can be borrowed or used on high school and university campuses when classes are not in session. If you think you will need facilities of this type, arrange for it before you commit too heavily to your idea. Stephen M. Jacobs told us he spent approximately $375 for equipment and supplies, but including borrowed equipment, the value of the project was about $13,000.

Caroline K. Horton, for her project on Fly Ash Foam, borrowed expensive equipment from companies and plants. "This," she said, "makes it possible to carry on research that would otherwise be impossible due to the cost factors."

Try to be as accurate and complete as possible when making your estimate. For many exhibitors we interviewed, the project ended up costing more than they had anticipated. The biggest exception to this was where the experiment was a continuation of a prior year's project, it was easier to budget the expenses. When you're making your estimate, include everything, from the supplies for the experiment to the materials for the graphs and the display board. Often, people forget that art supplies are sometimes the most expensive portion of the project.

In any event, making an estimate, and perhaps even budgeting or saving for these expenses, will help you avoid unpleasant surprises at the last minute.

One last step is to get the approval of your science teacher or advisor. Sometimes, you'll need him or her to sign a form, agreeing to the experiment.

Now that you've selected your topic, let's begin!

Fig. 2-12. Stephen Jacobs; *Design, Construction and Operation of a Continuous/Pulsed Ion Beam Accelerator* (close-up).

STUDENT PROPOSAL FOR SCIENCE PROJECT
MUST BE COMPLETED AND APPROVED BY TEACHER-ADVISOR BEFORE START OF EXPERIMENTATION

TO: _____ TEACHER

_____ HIGH SCHOOL

STATEMENT OF PROBLEM:

PROPOSED EXPERIMENTAL PROCEDURE: (Use back of sheet if necessary)

LOCATION WHERE EXPERIMENTAL PROCEDURES WILL BE DONE: _____

PROPOSED PROJECT CATEGORY: _____

I. **DOES EXPERIMENT INVOLVE LIVE, VERTEBRATE (NON-HUMAN) ANIMALS?** ☐YES ☐NO

IF YES — THE QUALIFIED ADULT WHO WILL SUPERVISE THE EXPERIMENT IN ACCORDANCE WITH THE REQUIREMENTS OF FORM GSDSEF-2 — "CERTIFICATION OF HUMANE TREATMENT OF LIVE VERTEBRATE ANIMALS"* IS:

NAME _____ DEGREE: _____ TITLE: _____

ADDRESS: _____ PHONE NUMBER _____

II. **DOES EXPERIMENT INVOLVE HUMAN SUBJECTS?** ☐YES ☐NO

IF YES — TO ENSURE THAT NO PHYSICAL, PSYCHOLOGICAL OR SOCIAL RISK IS INVOLVED, CAREFUL PLANNING WITH MY TEACHER-ADVISOR WILL INCLUDE STUDY AND COMPLETION OF FORM GSDSEF-3 — "CERTIFICATION OF COMPLIANCE OF RESEARCH INVOLVING HUMAN SUBJECTS."*

III. **DOES EXPERIMENT INVOLVE TISSUE SAMPLES OF HUMANS OR VEREBRATE ANIMALS?** ☐YES ☐NO

IF YES — THE INSTITUTION OR BIOMEDICAL SCIENTIST WHO WILL PROVIDE THE TISSUE IN ACCORDANCE WITH THE REQUIREMENTS OF FORM GSDSEF-4 — "CERTIFICATION OF TISSUE SAMPLE SOURCE"* — IS:

Name _____

Address _____ PHONE NUMBER _____

IV. **DOES EXPERIMENT INVOLVE RECOMBINANT DNA TECHNOLOGY?** ☐YES ☐NO

IF THE ANSWER TO EITHER I, II, OR III ABOVE IS **YES, I AGREE TO STUDY** AND **COMPLETE** THE NECESSARY CERTIFICATION, FORM GSDSEF-2 GSDSEF-3 or GSDSEF-4 **BEFORE** BEGINNING MY PROJECT.

IF THE ANSWER TO IV ABOVE IS YES, I WILL CONTACT ONE OF THE SCIENCE FAIR OFFICIALS LISTED IN THE STUDENT GUIDE TO ENSURE COMPLIANCE WITH REGULATIONS AND COMPLETION OF NECESSARY FORMS RELATED TO RECOMBINANT DNA RESEARCH.

_____   _____
Student Signature   Date

IN AGREEING TO SPONSOR THE STUDENT NAMED ABOVE, I CERTIFY THAT I HAVE REVIEWED THE PROPOSAL, THAT IT COMPLIES WITH THE RULES OF THE GREATER SAN DIEGO SCIENCE AND ENGINEERING FAIR, AND THAT, IF APPLICABLE, FORM GSDSEF-2, GSDSEF-3 or GSDSEF-4 HAS BEEN COMPLETED AND/OR THE STUDENT HAS COMPLIED WITH ALL THE ABOVE REQUIREMENTS.

_____   _____
Teacher-Advisor Signature   Date

*Form available from science teacher

Form GSDSEF-1 (1987)

Fig. 2-13. Student proposal for science project.

Chapter 3

Getting Organized

Now that you've chosen your topic, you are ready to begin organizing your project. Since this is probably the largest assignment you've ever done, the planning phase is the key to your success. Because of the length of time involved and the variety of activities you'll be conducting during the course of your science project, organization is crucial to this venture.

One of the biggest pitfalls in a long-term assignment is the tendency to leave things until the last minute (Take heart—this tendency to procrastinate is not limited to junior high and high school students—many people battle with this problem through their college years and into their careers!) It is, after all, easy to take advantage of a warm October day to ride your bike or play football instead of going to the library when the project isn't due until February!

If this project is required, your teacher has probably provided intermediate deadlines for you. This will help you to complete your project on time by giving you short-term goals. Even experienced students and many adults need milestones, with concrete results and items due, in order to keep going on a large project.

A schedule with short-term due dates will also help your teacher assist you, giving him or her the chance to help you correct small problems before they become insurmountable hurdles. Such a schedule may also help you identify problem areas, giving you the chance to change course while there is still time to find other resources.

If this project is not required, you may have to set up your own deadlines and schedules. Below is a "time line," one way of presenting a science project plan and schedule of due dates.

Because sheets like this tend to get lost, however, the schedule might be more useful if you transfer it to a large wall calendar and put it up in your room, where it will be a constant reminder.

You will notice "Log Check" occurring several times in this schedule. This is not a check of your Experimental Log, which is a diary of your experiment, and which we describe in Chapter 6. This Project log is a record of the entire project itself. Its purpose is to keep you organized and "on track." It is an informal record of all your activities on the project. This should include:

a. Research
b. Consulting the experts
c. Finding and buying supplies
d. Building equipment, if necessary
e. Setting up the experiment
f. Performing the experiment including testing and observation
g. Data analysis
h. Typing the paper
i. Building the backboard

The best way to keep a time log is to draw up a form, so that you can simply "fill in the blanks" as you work. Keep it available, so that you can make an entry every time you work on the project.

Even if this is not required, good records may prove useful later on. For example, one participant who needed data on solar activity contacted a division of the government agency, NOAA, who referred him to several U.S. and private agencies. One of these sources turned out to be a research institute library right in his own city! However, having the entire list might have been useful in case that resource did not turn out to be productive.

Documenting your activities could also be useful if you

```
         Timeline—Due Dates of Substeps
Date      Item
9/25      Topic Card, in the form of a question
10/4      Log Check
10/10     Bibliography Cards (at least 7 references)
10/20     Log Check
11/1      Notecards (at least 100, cross-referenced by source)
11/15     Log Check
11/20     First Draft - background research paper
11/30     Hypothesis, Procedures & materials list
12/1      Log Check
12/10     Revised hypothesis, Procedures & materials list (if
          necessary)
12/15     Background Research Paper (final)
12/20     Log Check
1/8       Log Check
1/17      Log Check
1/25      Log Check
2/5       Results (tables, graphs, etc.) and Conclusions
2/10      Project Notebook
2/20      Project Display
2/28      School Science Fair
3/31      City/County Science Fair
```

Fig. 3-1. Timeline schedule.

Table 3-1. Project Log.

Name_____

Project Title_____

Time Log

Date	Time	Activity
10/25/84	3:35 P.M.	Visited central library to find source material on eating habits of mice.
11/04/84	4:30 P.M.	Spoke to J. Jones, owner of Pet-a-rama pet shop to find out if 60 mice are available and how much the animals, cages and food would cost.
11/07/84	10:00 A.M. thru 2:00 P.M.	Spent time in library doing research for background research paper.

do not receive the responses you requested, or if delivery is late or non-existent on materials you ordered. This means not only making an entry in your log whenever you call or speak to anyone, but also keeping copies of any letters you send or receive in the course of your project. It will be easier to explain why you missed your deadline when you have documented evidence that you really did order your fruit flies back on November 1st, and called to inquire every Tuesday since that time!

Once you've drawn up a plan and a schedule, make sure that you let it be a tool to help you manage the project and not allow it to be something that rules you. If you find that the form you developed doesn't suit your needs, for example,

not allowing enough room to write all your notes, *change it!* Do not allow small annoyances to become large frustrations that prevent you from keeping accurate records.

Somewhere during the course of the project, if things aren't going the way you planned, show your teacher your records and explain the reasons you think the plan needs to be changed. He or she will be glad to help you work out the problem and advise you, but the task will be much easier if your documentation is good.

Remember that the purpose of organizing the project is to help you be successful, and not to restrict you to plans or procedures that are not working.

Chapter 4

Writing the
Background Research Paper

A background research paper is one of the required parts of a science project. For some of you, writing a paper of any kind is a more fearsome prospect than doing the experiment. You may imagine entire weekends spent unearthing dull facts. However, by breaking the process down into manageable steps, it can be less intimidating and more fun.

WHY DO A PAPER?

Before you begin, it will help to understand why the paper is necessary. There are three reasons that a research paper is required.

In order to conduct any worthwhile experimentation or research, it is essential to gather as much information as possible. This will help you to understand the theories, research and discoveries that have already taken place. It should provide a great deal of insight into your topic, and also save unnecessary steps in trying to prove some already established facts. Your research can also help you to narrow down your topic, particularly if the subject you've chosen is very general. This will be helpful later on, when you're trying to formulate your question and hypothesis.

Another reason applies to your careers as students. Throughout your education, countless papers of various lengths will be required in all subject areas. Some of these papers may be pure research, calling for gathering and organizing factual data. Others will be of a more interpretive nature, where you will need to learn the facts, and then show your understanding or opinion based on the knowledge you've gained. In any event, it's vital to know how to plan, organize and write the research paper.

This requirement also acquaints students with the actual process of conducting research. The ability to track down information is a skill you will use throughout your lives, even when your careers as students are over. As working adults, parents, or simply curious or informed citizens, the ability to find facts is essential. If your parents are helping you with this research, they'll know what I mean!

Actually, once you get started, this phase of your project can be lots of fun. In many ways, it's almost like playing detective. You may find one source that is not especially useful, but can offer a clue that may lead to three or four new sources. As with any other mystery, your research may lead you down some blind alleys, which will temporarily throw you off track. In the end, though, if you persevere, you'll certainly find what you are looking for. Along the way, you can learn many more fascinating things, perhaps on other subjects leading to new interests.

FINDING THE FACTS: A BRIEF GUIDE

The first step in the process is locating your information. A natural place to start is with the sources that are closest to you. These include your textbooks, any encyclopedias or other reference materials you may have at home, and publications in your school library. These references may yield only the most basic information, but will at least help you to focus on the various aspects of your topic. Also, the bibliographies in these volumes might point out other sources.

The next stop is your public library. If you do not know how to use the catalog there, ask the librarian's help. Local branch libraries often use card catalogs, employing the Dewey decimal system.

Other libraries use the LC, or Library of Congress, method to catalog books.

Both methods will guide you to the areas of the library

Dewey Decimal System		Library of Congress (LC) Classification System	
000	General Works	A	General Works
100	Philosophy	B	Philosophy, Psychology, Religion
200	Religion	C	Auxiliary Sciences of History (Archaeology, etc)
300	Social Science	D	History, General and Old World
400	Language	E, F	American History
	o 550 - The World Around Us	G	Geography, Anthropology and Recreation
	o 551.21 - Volcanoes	H	Social Science
500	Natural Science o 551.22 - Earthquakes	K	Law
	o 551.46 - Living Sea	L	Education
600	Useful Arts	M	Music and Books on Music
700	Fine Arts	N	Fine Arts
800	Literature	P	Language and Literature
900	History, Biography	Q	Science
		R	Medicine
		S	Agriculture
		T	Technology
	Ⓐ	U	Military Science
		V	Naval Science
		Z	Bibliography, Library Science
			Ⓑ

Fig. 4-1. Dewey decimal system (A) LC system (B).

where you'll find materials dealing with your subject.

Some libraries use a microfilm catalog in addition to the card catalog. This usually shows everything available in the entire library system. If the resources you need are in other branches, you may be able to order them delivered to your branch library. Otherwise, you'll need to travel to other parts of your city or county. Eventually, your search will probably lead you to your library's main branch to get all the information you need. If you live in a large city, this may be the opportunity to get acquainted with a sophisticated, computerized catalog system. This type of search will locate materials based on either author or subject.

Don't get the impression that you have to live in a big city to do research. To begin, of course, you'll have your school and local libraries. Then, all you need is writing materials and a post office, and you can gather material from all over the world. As you'll see later in this chapter, publications dealing with just about every conceivable area of science are available from the U.S. Government. Private companies are also cooperative, and quite willing to share information. Even foreign governments and international organizations might be willing to send material. The whole world's available for the price of a postage stamp!

For the moment, however, back to the library. Some materials may be available for you to check out. Before you bring the book home, scan it briefly to see that it really meets your needs. Check the copyright date to see if it's sufficiently recent to contain the most up-to-date information on your subject. For example, a book on nuclear energy written prior to 1950 would hardly be worth your time! In some rapidly expanding scientific fields, even information that is a few months old can be obsolete. Also, be sure that the material

is on your level. A manuscript written for use by a Ph.D. in biochemistry may be far beyond your capabilities if you're a seventh grader attempting your first science project. Looking at the table of contents, index and appendices might also give you a good idea of whether this book will be useful for you.

Other materials will be classified as reference material, which means you will have to use them at the library. If you need to do that, please remember library etiquette. Keep your materials close at hand and neat, talk in whispers, and no eating or drinking. Be sure you have plenty of paper and pencils, because there's usually no place to buy anything at the library.

Another way of doing research is to make copies of important information to bring home. Most libraries today have photocopiers, which are most helpful. But be prepared and bring plenty of change. These machines gobble up money faster than a video arcade!

Do not limit your research to books. There is a lot of valuable information in periodicals. This includes popular magazines, such as *Newsweek*, *Time*, and *Readers Digest*, and magazines that popularize scientific subjects, such as *National Geographic*, *Today's Health*, and *Psychology Today*. There are also specialized, scholarly periodicals that may deal with your subject. To find these types of articles, use a periodical index, such as *The Readers Guide to Periodic Literature*. Some of these indexes are:

—*Abstracts of Popular Culture*
—*Access*
—*Index to Free Periodicals*
—*Magazine Index*

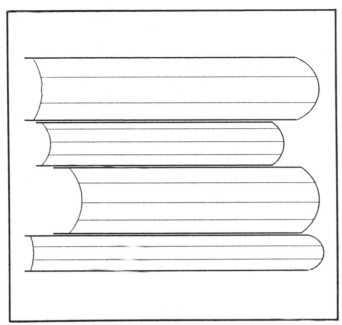

Fig. 4-2. Illustration of stack of books.

One of the most extensive sources of information is the U.S. Government, our nation's largest publisher. In fact, the various government agencies publish such a large variety of materials on a vast range of subjects, that in Lois Horowitz' book, *WHERE TO LOOK, the Ultimate Guide to Research,* the chapter on Government documents is the longest!

One way to locate helpful government references is to look at the list of U.S. government departments and agencies that publish extensively. Here is a list of just a few that publish scientific documents:

1. Department of Agriculture
2. Department of Commerce
3. Department of Defense
4. Department of Education
5. Department of Energy
6. Department of Health and Human Services
7. Department of the Interior
8. Department of State
9. Department of the Treasury
10. National Aeronautic and Space Administration
11. National Oceanographic and Atmospheric Administration
12. Bureau of Mines
13. Fish and Wildlife Service
14. Public Health Service
15. Atomic Energy Commission

Fig. 4-3. List of government sources.

Once you have located your sources, you'll find that if you require a recent edition, the library will usually have the actual issue on hand. Older issues will probably be available on microfilm or microfiche. Occasionally, the microfiche reader will have a copier attached (which also takes nickels, dimes, and quarters), so that you can make copies of important pages.

College and university libraries can also provide an invaluable resource for you. Often, they will have books and periodicals that a public library does not carry, especially if the college is noted for a particular area of science. The people there, who are sometimes student workers, can be very helpful in guiding you to the facilities available.

If a conveniently located university library has a great deal of useful information for you, investigate getting a library card. Many universities issue these cards for a fee, which is sometimes a tax deductible contribution. This will allow you the same library privileges as the college or university students, enabling you to check out nonreference materials.

While you're there, see if you can buy a copier card. This works something like a bus pass, because you prepay the cost of making a specific number of copies on the library's machine. This usually offers a saving over the cost of individual copies. It's also quite handy, because you don't have to worry about carrying a pocket full of change.

Institutes and foundations usually function the same way as university libraries. Without special access, their materials are only available for use on the premises, and cannot be taken out of the building (there are copiers here too)! However, these can be extremely good sources of specialized information. The information you find in institute libraries can perhaps lead you to experts in your field.

In your search for relevant government material, don't only look in the obvious places. The Department of Commerce, for example, publishes an annual entitled *United States Earthquakes,* and the Department of the Treasury, which is in charge of Alcohol and Tobacco, has many publications on those subjects.

There are several indexes to help you wade through the large amounts of government material. Several of these are:

1. *Monthly Catalog*
2. *Cumulative Subject Index*
3. *Publications Reference File*
4. *Index to U.S. Government Periodicals*

Even small libraries should have access to at least one of these research tools. The librarians may be able to help you use them. Once you have located the reference, it may not be readily available to you. Note, however, that some government publications can be purchased from the appropriate agency. Be sure you allow sufficient time to receive the materials you order. Many a project has faltered waiting for an order to be processed.

Branches of the military service, corporations and professional associations are also excellent sources of information on a variety of scientific and engineering topics. Often, they publish pamphlets which are yours for the asking!

Computer

If you have access to a large computer data base, such as *CompuServe* or *The Source*, it can help focus your research by providing a list of resources. However, be sure to plan your searches before you begin. Some of these time-sharing services charge different rates depending on the day of the week and time of day, much like using the telephone long distance. (Incidentally, don't forget about your phone charges when estimating the cost of using a large data base.) Also, because you're charged by the length of time you're connected, consider having the print-out mailed to you rather than checking the list on your screen.

As you look for your sources of information, review this checklist to be sure you've left no stone unturned!

TAKING NOTES

Now that you have located the information you need, you'll need to extract the most meaningful data to use in your background research paper. Because every fact you use will be derived from one of your sources, the first task is to start a working bibliography. Although this is a rather informal list, without regard for the "official" bibliography format you will use when submitting your final paper, include all the vital information about your source. Code each source with a letter or number, which you'll use later when writing up the index cards.

Table 4-1. Resource Checklist.

Places	Things
____School Library	____Encyclopedia
____Public Library	____Books
____University Library	____Magazines
____Institutes	____Scholarly journals
____Industry	____Professional journals
____Government	____Newspapers
____Military	____Government publications
____Zoos, Seaquariums	____Data Bases
____Museums	____Local newspaper

Examples of items in a working bibliography are as follows:

A. Encyclopedia International, Vol 6, Earthquakes
B. "Is California Sinking?," Scary Science Magazine, 1984

Fig. 4-4. Working bibliography.

You will, of course, want to take notes efficiently and effectively. The key to accomplishing this is to follow two important (and entirely contradictory) rules. First, focus on the most important and relevant facts, analyzing every piece of information, to determine whether or not it truly relates to your topic. Secondly, keep an open mind. Sometimes, a new, seemingly irrelevant fact can shed new light on your subject.

You will often find that information on several similar topics is included in one article or book. Extract only the information that furthers your understanding of the subject. If you find the same facts in several sources, use the presentation that makes the most sense to you.

Write each fact, in your own words, on an index card. It is important to use *your own words*. This will help you understand and absorb the material. However, if it's important to quote the material word-for-word, be sure to note that it's a direct quotation. Code each card to show the source of the fact and the page number.

Limit each index card to one fact only. This may seem like tedious work as well as a waste of space, but this method will make it a lot easier for you to write the final research paper. In addition, your science teacher may require you to turn in a specified number of index cards, along with your bibliography, as evidence of your research.

Computer Note:

If you have access to a computer, you may wish to use your word processor or data base software to create your index cards. In this way, your data may not have to be recopied when it's time to create your final research paper. Index cards are available for any tractor feed printer. Consult your software user manual to set up the print format for 3 × 5 cards.

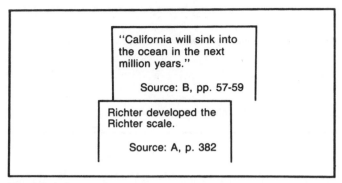

```
    "California will sink into
    the ocean in the next
    million years."

              Source: B, pp. 57-59

    Richter developed the
    Richter scale.

              Source: A, p. 382
```

Fig. 4-5. Index cards.

OUTLINING

Before you begin the actual writing, take the time to review all your facts. Try to determine a few themes, or main topics, which tie the material together. As you do this, some subtopics should also become apparent.

Write down, in outline form, what you want to say in your background research paper. If you are not well acquainted with using outline form, it is:

```
    I. Main topic I.

        A. Subtopic A
        B. Subtopic B
            1. Undersubtopic 1
            2. Undersubtopic 2
                a. sub-under-subtopic a
                b. sub-under-subtopic b
                    i. under-sub-under-sub-etc. i
                    ii. under-sub-under-sub-etc. ii
    II. Main topic II.
```

Fig. 4-6. Outline form.

Remember that in outline form, if an item is subdivided, it should contain *at least* two elements. Otherwise, the information should have been included in the division above. For example:

IV. Other insects
    A. Tarantulas
should read:

IV. Other insects; tarantulas, etc.

Once you've formulated the outline, you'll see how easy it is to arrange your facts. Use the outline to divide your index cards into stacks that relate to each topic and subtopic.

Don't worry if you have a few cards left over that do not seem to belong anywhere. They may either represent facts you've gathered that are irrelevant to your subject, or may point out a need to include another topic or subtopic in your outline. One helpful hint—you may want to keep

these extra cards although you don't want to include the facts in your background research paper. Data you may not need now could be extremely important as you progress in your experiment. Also, if you think towards the future, some science project participants build on their project year after year, going on to some new, original aspect each time. These facts that do not fit this year's experiment might be your guiding inspiration next year.

Next, arrange each stack of cards into a logical sequence. These ordered stacks of cards will now become the basis of the first draft of your research paper.

WRITING THE PAPER

Although your facts are now organized, let's take a moment to review the format of a well-constructed research paper. The first few paragraphs should introduce your paper, stating the topic you're writing about and the experiment you think you will do. Then, follow your outline to provide a logical presentation of your background information. The last several paragraphs should summarize the facts you've discussed, and finally, you should predict what you believe your experiment will prove or disprove.

Note: Computer

If you have recorded your facts using a data base or a word processor, use your software to rearrange your facts in the same sequence as you've classified your index cards. Delete any unrelated facts and store your file under a *new name.* This will preserve a copy of your old list "just in case."

If you've recorded all the facts you've found in your own words, it will be easier to create a research paper that is truly your own. Remember, however, that the paper does not simply consist of a bunch of facts strung together. Your teachers, as well as science fair judges, will look for some creative thought and educated interpretation of the facts, which tie your research to the experiment you will conduct.

One resource you'll find valuable as you begin to write is Strunk & White's *Elements of Style.* Originally published in 1919, it is the definitive guide to good, clear writing. This small, inexpensive paperback is one you will use frequently throughout your student career. The book not only describes rules of correct grammar, but also gives guidelines for writing clearly and concisely. Especially for inexperienced

1. For the first footnoted reference to a book:
   [1]Robert W. McLuggage, *A History of the Americal Dental Association* (Chicago, American Dental Association, 1959), p. 475
2. For subsequent footnoted references:
   [4]McLuggage, pp 15-18
3. For a periodical
   [12]*Psychology Today*, October 1986, 68

(NOTE: In all cases, the materials cited in your footnotes should appear in your bibliography, in proper format, as well.)

Fig. 4-7. Footnote format.

writers, the examples of misused, overworked, and common, incorrect expressions might be the book's most helpful feature. Following even a fraction of the helpful hints in *Elements of Style* will help you create a more polished and readable research paper.

CREDIT YOUR SOURCES

At this point, we must discuss an unpleasant little topic, called plagiarism. It's a topic you'll hear discussed throughout your educational career. Webster's Unabridged Dictionary defines the verb *plagiarize* as "to take and pass off as one's own (the ideas, writings, etc., of another)." To put it simply, if you use an author's exact words, be sure to put them in quotes, for example, if Ms. XY says in her book entitled *Flea Flicking Made Simple*, "the most efficient way to flick fleas is using the thumb and forefinger." If you don't use an exact quote but very closely paraphrase the author, you could say: Ms. XY tells us, in *Flea Flicking Made Simple* that using the thumb and forefinger is a most effective method.

Another way of crediting the author is by using a footnote. You could state in your paper: Flea flicking is best accomplished with the thumb and forefinger[1], where a footnote, numbered 1, citing Ms. XY's book would appear on the bottom of the page.

As you progress in the academic world, plagiarism is considered an extremely serious offense, since it is, in a sense, taking someone else's original ideas or research and claiming it for your own. Therefore, be sure to get into good habits of crediting your sources, right at the start.

Just as with the bibliography, footnotes follow a specified format.

WRITE AND REWRITE

Your rough draft is just that, a first attempt at writing your research paper. There are no hard and fast rules as to how many drafts a paper should go through, but this is the opportunity to correct, revise, rearrange and reword your material to its best advantage. You may wish to type or print your early drafts with triple spacing, to give yourself enough room to write in many corrections.

Have family members and friends review your work too. Although they may not be scientists or writers, they can find spelling errors, awkward grammar, or badly worded sentences that you are too close to spot. Keep your handy dictionary and thesaurus at your side as you write.

A. Periodical
   Patrick Huygne, "Earthquakes, the Solar Connection" *Science Digest,* XC, (October, 1982), 73-75
B. Interview
   Dr. J. Holman III, Orthodontist, *Interview,* 11/23/85
C. Books
   1) by single author:
      Lauber, Patricia, *Of Man and Mouse.* New York: The Viking Press, 1971
   2) by multiple authors:
      Anderson, Garron P., S. John Bennett and K. Lawrence DeVries, *Analysis and Testing of Adhesive Bonds.* Long Beach, CA: Foster Publishing Co., 1971

D. Encyclopedias
   1) Selection without author
      "Galena," *Encyclopedia International,* 1974
   2) Selection with author
      Roderick, Thomas H., "Gene," *Encyclopedia International,* 1974

E. Government or other institutional publication
   Science Service. *ABSTRACTS - 37th Internation Science and Engineering Fair,* Washington, D.C., 1986

Fig. 4-8. Final bibliography.

a good impression on science fair judges.

Regardless of how it was created, be sure your margins are correct, your text is double-spaced, your name is on the paper and you have obeyed any formatting rules your teachers have given you. Remember to include your bibliography, in finalized format, at the end of the report.

Computer Note:

This may be almost too obvious to mention. Those of you with computers can throw away the erasers, correction tape, and white fluid, and just use your keyboard.

If you do not have a spelling checker, this may be the time to get one. These programs are quite inexpensive, and will give you valuable assistance throughout your academic career. Some of them even include a thesaurus, which will help you to find synonyms for overworked words.

When you are satisfied that your paper is the best that it can be, you are ready to create your final copy. If, until now, your drafts have been handwritten, try to borrow a typewriter. Even if your handwriting is good, your teacher will appreciate seeing typed copy, and it will certainly make

Computer Note:

If you have been using a computer, this, of course, is simply printing out your latest version of the report. Now that this phase of the project is complete, remember to make a backup diskette. Be sure that all your files, the notes, bibliography, and footnotes, as well as the final research paper, are included. This is your insurance policy!

Chapter 5

The Question and the Hypothesis

"Why is there air?" asks Bill Cosby, at the beginning of one of his famous routines. Although you would hardly ask such a broad or general question as the basis of a science project, (especially since you could never hope to finish the experiment in time!) the objective of *all* science is to answer a question or solve a mystery concerning the world around us.

The experiment you'll do, however, should try to solve a very specific problem. When you are deciding on the question (or purpose of the experiment) be sure that it's one that you can answer, with the time, resources, skills, and equipment available to you. It therefore must zero in on a limited aspect of your general topic. When focusing on your main purpose, keep in mind exactly how you plan to carry out your experiment and what you need to do it. After you consider these factors you may change the question you've selected.

The information you gathered while you were writing your research paper should come in handy now, helping you to focus on the purpose of the project, and pose that specific question. Remember, however, that an effective project cannot be based on an inquiry that can be answered with research alone. Your problem should be one you can solve by experimentation, followed by analysis of the results.

Notice that we've used the word *specific* several times. The biggest mistake beginners make at this point in the project is to define their topic too broadly.

You will notice that the questions in Fig. 5-1 are very general. They are vague, and do not indicate the substances or quantities to be tested, or how differences will be measured.

Instead, formulate your question, identifying exactly what your experiment will test and the comparisons you

intend to make. Examine Fig. 5-2 and notice that the same topics used above have been narrowed down and reworded into effective questions that express the problem more specifically.

This set of questions indicates what will be compared and what measurements will be used to show similarities or differences. For example, question 3 not only states what we are attempting to accomplish with the experiment, but also what is to be measured.

When you carefully read and compare all five questions on both lists, you'll see that the best questions for a science project are those that are extremely limited in scope. They state the intent of the experiment, list the subjects to be tested, and then describe results to be measured.

Once you have decided on your question, you must develop your *hypothesis* before designing and executing your experiment. You probably have a good guess, or at least an opinion, on what you think the answer to your question will be. You may have held this theory before doing your research, but after gathering the data, this should be an educated guess, based on the information you've learned so far.

Let's stop here for a moment now, and be sure that we understand all the terms. A *fact* is something that has been proven true. An *opinion* is something you believe to be true. Webster's Ninth New Collegiate Dictionary defines an opinion as: "a belief stronger than an impression and less strong than positive knowledge." The hypothesis, on the other hand, is more than an opinion and less than a fact. It represents your prediction of what the results will be, or, in other words, your *opinion* based on the *facts* that you've learned.

The experiment will be a test, or series of trials, which

1. Does television have a bad affect on test scores?
2. Does sound affect appetite?
3. What plants are flame retardant?
4. What is the effect of irrigating plants with salt water?
5. Do teenagers get overexcited watching MTV?

Fig. 5-1. Poorly worded questions.

1. What is the difference in test scores between sixth graders watching less than 3 hours of TV per night and 3 or more hours of TV per night?
2. What is the difference between the appetite in mice subjected to high and low frequency sounds, or no sound at all?
3. What are the rates of combustion of five different types of plants growing in the canyon areas of San Diego?
4. What is the difference in growth rate of three different types of plants irrigated with tap water and the same plants irrigated with sea water?
5. What is the difference in pulse rate between eighth graders watching MTV and listening to classical music?

Fig. 5-2. Well worded questions.

1. I believe that the test scores of 6th graders who watch more than 3 hours of TV per evening will be at least 20% lower than those watching less than 3 hours, because children watching more TV spend less time reading.
2. I believe that the amount of food eaten by mice subjected to high and low frequency sounds will differ from the amount eaten by those not subjected to those sounds.
3. I believe that iceplant and other succulents burn more slowly than other canyon plants, such as mesquite, sage, and giant dandelions, due to the greater degree of stored moisture.
4. I believe that sea water will retard growth in house plants such as spider plants, pothos, and Boston Fern, due to the high salt content, which is toxic to plants.
5. I believe the pulse rate in eighth graders watching MTV will be 50% higher than the pulse rate of those listening to classical music, because MTV is louder and faster.

Fig. 5-3. Hypotheses.

you will perform in order to prove your theory. You will apply various substances or use different quantities of something on a particular subject, then measure the variations, and finally record results. Therefore, your statement must identify *what* you are testing, *how* you are testing it and *why* you expect the results to prove your hypothesis. Figure 5-3 gives examples of well-worded hypotheses that relate to the questions asked in Fig. 5-2.

You'll notice that all the hypotheses begin with the words "I believe . . ." This indicates that it is only a theory until the experiment has been done and its results analyzed.

Sometimes a hypothesis will contain not only your theory of what the experiment will prove, but why you're making this assumption. In Andrew Wolf's experiment testing the porosity of various types of Hohokam pottery, he assumed that cooking required pottery of different porosity than storage.

In certain types of projects, the hypothesis is not so much a theory as a statement of what the experiment hopes to accomplish. In William Chen's project BLACKJACK, his work did not attempt to prove a theory, but instead, tried to improve different blackjack systems available. When Angel de la Cruz developed his computer program

Mendelevia, he was not attempting to prove a hypothesis, but rather to create a computer game that would make a boring subject fun to learn.

Although you will not need a complete list of variables and controls at this time, you will need at least a rough idea of these in order to formulate your hypothesis. Briefly, a variable is something to be changed during the experiment and a control is something to be held constant, against which measurement is made. As shown in the examples, your hypothesis should mention:

The subject of your experiment
—The variable to be changed
—The variable to be measured
—The results you expect

The subjects of the first experiment are 6th grade students. The measured variable is the test scores, and the experimental variable is the amount of television watched. The result expected is that the students watching more TV will have lower scores. In the third example listed, the subject to be tested is brushfire, the variable to be tested is the type of plant, the variable to be measured is the rate of burning, and the results you expect are that succulent plants burn more slowly than nonsucculent plants. In the fourth example, the variable to be changed is the type of water used, the controls are the type of plant and amount of water, and the measured variable is the amount of growth.

Although the Question and Hypothesis are the shortest pieces of writing you'll do during your project, they're vitally important. They form the basis of the entire experiment, and are the basis on which your results are judged.

Chapter 6

The Experiment

The heart of your science project is the experiment. To be effective and successful, the experiment must be carefully and methodically planned. Science fair judges on all levels overwhelmingly agree that the most important factor in rating your entire project is the proper use of the scientific method.

The scientific method is an organized system of conducting an experiment, including the collection, measurement and documentation of data. It must include a specified variable and control, as well as at least one experimental and one control group. In addition, the experiment must use a large enough number of samples, or perform sufficient repetitions of the test, to assure a valid result.

Let's begin by defining the terms *subject, variable*, and *control,* and how they are used.

SUBJECT OF THE EXPERIMENT

Any valid experiment is designed to test and examine the effect of a change in environment or condition. To conduct such a test, there must be a specific subject, one or more variables, and a control. The subject of the experiment is the item being tested. For example, in John Stoffel's project, *The Effect of Radiation on Bean Seeds and Seedlings,* the bean seeds and seedlings are the subjects of the experiment.

The rabbits are the subject of Elizabeth M. Robinson's project entitled *The Rabbit Test,* in which she tested the effects of steroids on the reproductive rates of rabbits.

VARIABLES

One variable in your experiment is the factor that is altered. This is often the substance or condition that your project is studying. For example in Nicola Kean's project *What happens when the length of a Pendulum is increased?* the length of the pendulum is the variable. In an experiment that concerned radiation and plants, conducted by Jeffrey S. Morgenheim, the amount of laser radiation was the variable. These are examples of what is called either the *experimental variable,* or the *independent variable.*

Another type of variable is the *measured variable,* which is also known as the *dependent variable.* This represents what you are evaluating, such as the amount of growth in a particular specimen. Daniel S. Chang, of Pago Pago, American Samoa conducted an experiment to determine whether carbon dioxide was necessary for coral growth. The amount of growth in his coral specimens was the measured variable.

Let's examine some of the projects we discussed in Chapter 5 and look at the subject as well as the independent and dependent (or experimental and measured) variables shown in Table 6-1.

CONTROLS

Once the variables have been determined, it's important to list the things that must not be allowed to change. These are known as the *controls,* or *controlled variables.* (Don't confuse this term with "control group," which we'll discuss in a few minutes.) Controls are applied to *all* subjects being tested. If your experiment changes in any way, other than the stated variables, it will be impossible to determine whether your intended variable or other dissimilar factors caused the results that occurred. For example, if testing the effects of caffeine on different groups of goldfish, the amount and type of food given to the groups, as well as the lighting

Fig. 6-1. John Stoffel; *The Effect of Radiation on Bean Seeds and Seedlings.*

available to the fish in the separate tanks, must be identical. Otherwise, it would be impossible to prove whether differences in the groups at the end of the experiment were caused by varying amounts of caffeine, or the alterations in diet, lighting, etc. Notice the controls for our famous five projects shown in Table 6-2 will keep all the factors except the independent variable identical throughout the experiment.

EXPERIMENTAL AND CONTROL GROUPS

In order to make a statement about the results of your experiment, you must compare the results of applying your variable to your subject with your subject under normal conditions. In other words, what happens in the course of your experiment must be compared to the condition of your subject before anything was changed. To accomplish this you will need to divide your subjects into *at least two groups.*

Your *experimental group* is a collection of subjects to which the *experimental variable* is applied. If you have multiple independent variables, apply one at a time. The reason for this is that if you change several variables simultaneously, you'll never be sure exactly what caused your results. Table 6-3 lists experimental and control groups.

It is also very important to be sure that all the subjects used in both groups are identical. For example, in Daniel S. Chang's project, which determined whether carbon dioxide was necessary for coral growth, the most difficult part of the project was collecting identical coral specimens for the experimental and control groups.

Another way of applying multiple variables, or several forms of the same variable, in order to carry out the purpose

Table 6-1. Subjects and Variables.

Subject	Experimental Variable	Measured Variable
Sixth graders	Hours of TV watched	Test scores
Mice	Type of sound	Appetite
Brush fire	Species of plant	Rate of combustion
House plants	Type of water	Amount of growth
Eighth graders	Variety of music heard	Heart rate

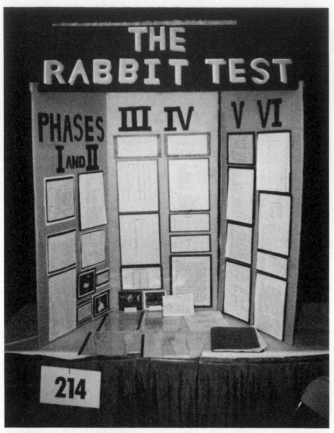

Fig. 6-2. Elizabeth M. Robinson; *The Rabbit Test; Phases V and VI.*

of your experiment, is to use a number of experimental groups. In Randall Shank's experiment, *NutraSweet ™ In Diet Soda: A Study of Storage Temperature* there were four experimental groups, stored at different temperatures, 5°, 15°, 25°, and 35° C. In another project studying popular soft drinks, Jennifer L. Decker tested the sugar content in eight different regular and diet soft drinks, using glucose solutions and testing for the presence of sugar.

To make your experiment manageable, restrict the number of variables or variations. In other words, if you limit the factors you're changing and hence, the number of experimental groups you will use, you'll keep your squash plant experiment from being the killer gourd that ate New Jersey.

The *control group* is a collection of subjects, identical to those in the experimental groups studied, to which no variables are applied. For example, if you are testing the effect of crushed bone meal added to the soil for tomato plants, your control group would consist of a group of tomato plants that *did not have* bone meal added to the soil. With the exception of not applying the stated variable to the control group, it must be *exactly* like the other group(s) in all other respects, such as type, number, environment, brand, etc. Otherwise, the comparison, and hence, the results of your experiment, may be invalid.

For example, in Libby Haines' project, *Can A Charcoal Filter Reduce the Effects of Automobile Exhausts on Chick Embryos?* the control group consisted of embryos which were not subjected to any fumes, as opposed to the experimental groups, which were subjected to filtered and unfiltered fumes.

PROCEDURES

The first step in designing your project is to detail the exact experimental procedures. It should be a step-by-step list that anyone, even a person who knows nothing about science, could follow in order to duplicate the experiment. Each item in the list should describe only one action, in the exact sequence that the particular step is performed in the course of the experiment. Because you are breaking the

Table 6-2. Project Controls.

Subject	Project Controls
Sixth graders	The type of courses taken, the other environmental factors, such as whether they have their own room.
Mice	Amount of light and water each group gets.
Brush fire	Canyon areas where plants are native.
House plants	Amount of light and plant food each specimen receives.
Eighth graders	Amount of aerobic and other activities the Students indulge in.

Table 6-3. Experimental and Control Groups.

Subject	Experimental Group(s)	Control Group
Sixth graders	Those that watch more than 3 hours daily	Those that watch less than 3 hours daily
Mice	Those that are subject to periods sound	Those not subjected to periods of sound
Brush fire	Burning of succulent canyon plants	Burning of non-succulent canyon plants
House plants	Group watered with sea water	Group watered with tap water
Eighth graders	Group listening to MTV	Group listening to classical music

---

Procedures

1. Buy 30 mice.
2. Buy cages and food.
3. Put 10 mice in each cage and put them in separate rooms.
4. Give mice the same amount of food each day.
5. Test two of the groups with sound.
6. Measure and record the amount of food that each group eats.

Fig. 6-3. Poor procedures.

procedure down to such thorough detail, it may contain several pages, including not only the experiment itself, but the steps taken to buy, borrow, or build your materials, set up the experiment, and finally, conduct the necessary tests.

Probably, your advisor or science teacher will want to review your procedures before you actually begin. He or she will examine them very critically, looking for specific, detailed, sequential steps. Next, the teacher will review each step. If your sample is too small, or you plan to conduct too few tests, he or she may ask you to add steps or describe them in more detail.

"But I meant that," you might protest. "Anyway, that's really included in step 4." It may be obvious to you, but would probably be a complete mystery to that theoretical nonscientist who should be able to duplicate your experiment.

Study Fig. 6-3. It shows a list of procedures for a well thought-out experiment involving the correlation between sound and mouse appetite. However, the steps are too vague for anyone who's uninformed on the topic.

However, by simply spelling out every step in detail, as shown in Fig. 6-4, the list becomes an effective statement of procedures.

If you are using live vertebrate specimens, such as mice, show the steps you'll be taking to insure the adequate care and humane treatment of your subjects. Get the necessary permissions of parents, teachers, or other authorities.

Also, if your project uses tissues, organs, human, or animal parts, you will need to complete a form certifying the source of your tissue sample.

MATERIALS

Not only must the steps be precise, but they must include a materials inventory, which enumerates *everything* you will use. Again, this should be a list that is so specific that anyone could follow it. List the brands, sizes, quantities, contents and temperatures of all items that are necessary to conducting the experiment. Figure 6-7 compares vague and specific materials lists for the same project.

If there is anything you are using that you cannot describe verbally, illustrate it by including diagrams or photographs, making certain to include specific details of each item.

Some science students build their own machines or containers to use in their experiments. If you've done so, include instructions on how to construct it. These directions should be similar to (or, hopefully, better than) those that are included when buying a bicycle or a barbecue that requires assembly. They should include a parts list, as well as a description (and perhaps an illustration or photograph) of the finished product. Here too, be sure to list *specifically* all materials used. The instructions should consist of a numbered, sequential list, detailing what parts are used, in which order, to construct the device. Rather than something "so simple that even a child of four could do it," strive for a set of instructions that are easy enough for an ordinary parent to follow!

Procedures

*1. Buy 30 mice, 3 cage kits, rodent bedding material and mouse food from ABC Pet Shop.
*2. Buy a high frequency buzzer and a low frequency buzzer.
3. Set up each cage in a different room of the house.
4. Put 88 cubic centimeters of bedding, 1.70 cubic centimeters of food and 60 cubic centimeters of water in each cage.
5. Put 5 male and 5 female mice in each cage, and label the cages A, B and C.
6. Wait 3 days to begin experiment to allow mice to get adjusted to surroundings. During that period, measure amount of food and water consumed.
7. Each day during experimental period (11/18 through 2/2):
   a. In each cage, measure the amount of food left in the food dish. Subtract the amount from the amount given each morning (1.87 cubic centimeters) to determine the amount eaten, which is recorded in the experimental log.
   b. Fill food dish in each cage with 1.87 cubic centimeters each morning. (The amount consumed during pre-test periods plus 10%)
   c. Refill water bottle in each cage.
   d. Each day, expose the mice in group B to the low frequency sound for 1 hour and expose the mice in group C to the high frequency sound for 1 hour.
8. Each Saturday during the experimental period, move the mice in each cage to a large cardboard box (together with their food dish) while cages are cleaned and the cedar bedding is replaced.
9. When the experiment is completed, re-sell the cages and mice back to ABC Pet Shop.
*For details, see materials list.

Fig. 6-4. Good procedures.

CERTIFICATION OF HUMANE TREATMENT OF LIVE VERTEBRATE ANIMALS

Name of Entrant _____ School _____

Project Title _____

Any student research involving animals **MUST COMPLY** with the requirements of the **California Education Code** stated here:

HUMANE TREATMENT OF ANIMALS, State of California Education Code Title 2, Division 2, Part 28, Chapter 4, Article 5

51540. In the public elementary and high schools or in public elementary and high school sponsored activities and classes held elsewhere than on school premises, live vertebrate animals shall not, as part of a scientific experiment or any purpose whatever:

      (a) Be experimentally medicated or drugged in a manner to cause painful reactions or induce painful or lethal pathological conditions.

      (b) Be injured through any other treatments, including, but not limited to, anesthetization or electric shock.

Live animals on the premises of a public elementary or high school shall be housed and cared for in a humane and safe manner.

The provisions of this section are not intended to prohibit or constrain vocational instruction in the normal practice of animal husbandry.

Compliance with the following ISEF regulations is **ALSO REQUIRED**; however, **THE PROVISIONS OF THE CALIFORNIA EDUCATION CODE MUST BE FOLLOWED WHENEVER CONFLICTING REGULATIONS OCCUR.**

EXCERPTS, INTERNATIONAL SCIENCE AND ENGINEERING FAIR (ISEF) REGULATIONS FOR EXPERIMENTS WITH ANIMALS

Prepared by Science Service, Inc., for exhibitors at the ISEF and Affiliated Fairs

1. The basic aims of experiments involving animals are to achieve an understanding of life processes and to further knowledge. They do not include the development of new or refinement of existing surgical techniques or experiments in toxicological studies. Experiments involving animals (live or preserved, vertebrate or invertebrate excluding **Homo sapiens**), vertebrated embryos and fetuses, and embryos of fowl within three days of hatching, must have clearly defined objectives requiring the use of animals to demonstrate a biological principal or answer scientific propositions. Such experiments **must** be conducted with a respect for life and an appreciation of humane considerations.

2. The use of protista and other invertebrates is to be encouraged for most research involving animals. Their wide variety and the feasibility of using larger numbers than is usually possible with vertebrates makes them especially suitable.

3. To provide for humane treatment of animals, an animal care supervisor knowledgeable in the proper care and handling of experimental animals **must** assume primary responsibility for the conditions under which the animals are maintained. If the school faculty includes no one with adequate training in this area, the services of a qualified consultant **must** be obtained.

4. All live or preserved animals or animal parts **must** be lawfully acquired from an approved source and their care and use **must** be in compliance with local, state and federal laws.

5. The **comfort** of the animals shall be a prime concern. No research using live vertebrate animals shall be attempted unless the animals are obtained from a reliable source and the following conditions can be assured: appropriate, comfortable quarters; adequate food and water; humane treatment and gentle handling. Care must be provided at all times, including weekends and vacation periods.

Fig. 6-5. Certification of humane treatment of live vertebrate animals.

GSDSEF ANIMAL REGULATIONS

1. Experiments involving procedures not in violation of the "painful reaction" or "injured" restrictions of the California Education Code are permitted if certified by a qualified biomedical scientist PRIOR TO the beginning of the investigation. (NOTE: GSDSEF rules do not permit students or their adult supervisors, as part of a student-planned project, to 1) perform surgery; 2) conduct experiments involving toxicity, nutritional deficiency or harmful physical or psychological stress or 3) perform the sacrifice of live vertebrate animals.

2. It is permissible for the student and designated adult supervisor to consult with a biomedical scientist to obtain detailed instructions and guidance in techniques to be used by the student under the direct continuous supervision of a designated adult supervisor (for research not conducted in the biomedical scientist's lab). In this instance the designated adult supervisor will be required to certify in writing jointly with the biomedical scientist.

3. Either the biomedical scientist or adult supervisor must provide continuing supervision to assure compliance with the protocol.

4. Major deviations from the approved protocol may be implemented only with the written approval of the biomedical scientist.

5. The biomedical scientist or adult supervisor must be in the same locality as the student for the duration of the experimental work except for short trips. This means that a project started in one city may not be continued in another unless an alternate designated adult supervisor, approved by the biomedical scientist prior to the continuation of the experimental work, agrees to supervise the project.

6. A biomedical scientist is defined as one who possesses an earned doctoral degree in science or medicine and who has current working knowledge of the techniques to be used in the research under consideration.

7. A designated adult supervisor is defined as an individual who has been properly trained in the techniques and procedures to be used in the investigation. The biomedical scientist must certify that the designated adult supervisor has been so trained.

RESEARCH PLAN

Purpose of Project:

Starting Date: _____

Site at which investigation will take place:

    Name _____

    Address _____

Live vertebrate animals to be used:

    a) Genus, species and common name _____

    b) Number of animals _____

    c) Animals obtained from _____

List objectives of the experiment and describe fully the methods and techniques involved. When the use of electrical current, laser beams, sound stimuli or other artificial stimuli are an integral part of the Research Plan, they must not exceed the normal tissue tolerances for the species concerned (AS INDICATED IN THE **Biology Data Handbook,** 2nd Edition; editors, P.O. Altman and D.S. Dittmer; publisher, Federation of American Societies for Experimental Biology).

Describe proposed methods of animal care:

    a) Cage size _____

    b) Number of animals per cage _____

c) Temperature range (maximum and minimum) in degrees Celcius of room where animals are to be kept.

_____

d) Frequency of feeding and watering _____

e) Frequency of cleaning cage _____

f) Type of bedding to be used _____

g) Where will animals be housed? _____

h) Where will animals be returned when research is complete?

_____

Name of Animal Care Supervisor _____ Name of Biomedical Scientist _____

Name of Designated Adult Supervisor _____

Signature of Student _____

CERTIFICATIONS
THE FIRST TWO CERTIFICATIONS MUST BE COMPLETED FOR ALL PROJECTS INVOLVING LIVE VERTEBRATE ANIMALS

CERTIFICATION BY TEACHER/ADVISOR   I agree to sponsor the student named above and assume responsibility for compliance with the existing rules and regulations pertaining to experiments with animals.

Signature _____

Name (type or print) _____   Date _____

Institution _____   Title _____

Institution Address _____   Phone _____

Home Address _____   Home Phone _____

CERTIFICATION BY ANIMAL CARE SUPERVISOR   of compliance with California Education Code and Paragraph 3 of the ISEF Regulations for Experiments with Animals. (Must be completed prior to receipt of animals by the student.)

I certify that I have reviewed and approved the Research Plan and will supervise and accept primary responsibility for the quality of care and handling of the live vertebrate animals used by the designated student. I further certify that I am knowledgeable in the proper care and handling of experimental animals and meet prevailing animal supervisory requirements.

Signature _____

Name (type or print) _____   Date _____

Institution _____   Title _____

Institution Address _____   Phone _____

Home Address _____   Home Phone _____

This section MUST be completed for ALL vertebrate animal projects.

Fig. 6-5. Continued.

Complete this section for all projects except observations of animals in their natural environment.

CERTIFICATION BY BIOMEDICAL SCIENTIST (if required) of compliance with the California Education Code, the ISEF Regulations for Experiments with Animals and the GSDSEF Animal Regulations. (Must be completed prior to the start of the project.)

I certify that I have reviewed and approved the Research Plan; that if the student or designated adult supervisor is not trained in the necessary procedures I will ensure his/her training; that I will assure that the requirements of the California Education Code, the ISEF Regulations for Experiments with Animals and the GSDSEF Animal Regulations are fully met; that I will provide advice and supervision personally or through a designated adult supervisor throughout the project; and that I am a qualified scientist with an earned doctoral degree (Ph.D., M.D., D.V.M.) and a working knowledge of the techniques to be used by the student in this research.

Signature _____

Name (type or print) _____ Date _____

Institution _____ Title _____

Institution Address _____ Phone _____

Home Address _____ Home Phone _____

CERTIFICATION BY DESIGNATED ADULT SUPERVISOR (if required)

I certify that I have been trained in the techniques to be used by this student and will provide direct supervision for the research.

Signature _____

Name (type or print) _____ Date _____

Institution _____ Title _____

Institution Address _____ Phone _____

Home Address _____ Home Phone _____

This form, properly completed, must be part of the carefully planned procedures for experimentation with live vertebrate animals or animal parts. It must accompany any such project exhibited at, or when presented for any public display associated with the Greater San Diego Science and Engineering Fair. All sections must be completed except the "Certification by Biomedical Scientist" and the "Certification by Designated Adult Supervisor" which need be completed only by those students carrying on experiments other than observations of animals in their natural environment.

Form GSDSEF-2 (1987)

CERTIFICATION OF TISSUE SAMPLE SOURCE

Name of Entrant _____ School _____

Project Title _____

This form must be completed for all projects using tissue(s), organ(s), human part(s), or animal parts, including blood.

When live or preserved tissue samples or parts of human or vertebrate animals are obtained by the student from an institution or biomedical scientist, a statement signed by the adult providing the tissue is required. Students may NOT be involved in the direct acquisition of these samples from living human or vertebrate animals.

Live tissue samples must be:

    a) from a continuously maintained tissue culture line already available to institutional researchers.
       or
    b) from animals already being used in an on-going institutional research project.

RESEARCH PLAN

1. Tissue(s), organ(s), or part(s) used: _____
                                   _____
                                   _____

    Tissue sample is from: (check appropriate box)
    ☐ Human Source
    ☐ Vertebrate animal source
    Genus, species and common name _____
    _____

2. Starting Date: _____

3. Purpose of Project:

Fig. 6-6. Certification of tissue sample source.

4. List objectives of the experiment and describe fully the methods and techniques involved:

Signature of Student _____ Date _____

CERTIFICATION

Institution or company that is source of Tissue Sample:

Name _____

Address: _____

_____

I certify that the above listed materials were provided by me or my institution and that the student listed was NOT involved in the direct acquisition of the samples provided or purchased.

_____   _____
Signature   Title

_____   _____
Date   Telephone

Form GSDSEF-4 (1987)

Vague		Specific	
1.	30 mice	1.	30 house mice, 15 of each sex.
2.	2 buzzers	2.	One piezzo sound buzzer of high frequency (3800 cycles per second).
3.	3 cage kits	3.	One piezzo sound buzzer of low frequency (1500 cycles per second).
4.	Mouse food	4.	Three identical cages measuring 11 × 16 × 2 inches each.
5.	Cedar bedding	5.	Three identical play wheels.
6.	Food dishes	6.	7 kg. mouse food, consisting of corn, sunflower seeds, wheat and oat grain.
7.	Water bottles	7.	975 cubic centimeters of cedar chip bedding.
		8.	Three identical water bottles.
		9.	Three identical food dishes, each with a volume of 3.4 cc.

Fig. 6-7. Materials list.

Many extraordinary projects can be done using very ordinary materials. However, if you require special equipment or supplies, you may need some help in locating these, or else you may need to alter your experimental design. Before deciding to build or purchase what you need, find out if you can borrow the equipment from your school laboratory. Teachers may permit you to use it after school hours, or even allow you to take some items home. However, first investigate what the fees are for any loss or breakage. You may have second thoughts about using these things without supervision.

Other possibilities are university or corporate lab facilities. This has been a particularly good resource for those with a mentor or advisor who works in a laboratory. Several students with successful projects have developed excellent working relationships with scientists who have helped them. Sometimes, friendships have developed that extended far beyond the life of the science project.

It may turn out that your only course of action is to buy your materials. If so, see if you can buy these through your school in order to take advantage of discounts that may be available to a large school district. If not, you will have to contact various supply houses for the live specimens, chemicals, or equipment you may need. The best place to begin is your local *Yellow Pages,* for Chemicals, Biological Supplies, Laboratory Equipment, Plant Nurseries, etc.

If your experiment is concerned with rare, or exotic

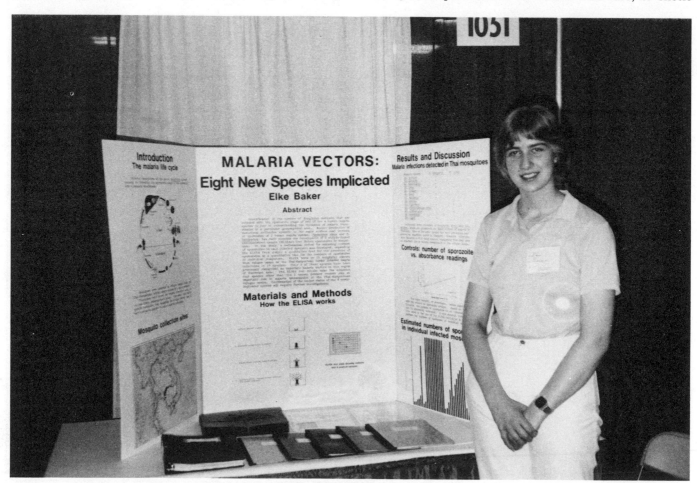

Fig. 6-8. Elke Baker; *Malaria Vectors—Eight New Species Implicated.*

Fig. 6-9. Lan Do; *The Influence of Favorite Color, Sex, Age, Eye and Hair Color upon Wardrobe Color.*

Table 6-4. Sample Size.

Subjects	Suggested number of subjects per group
Plants	10-50
Live Vertebrates	10-15
Humans	50-100

specimens or substances, your search for the proper materials will be even wider. Elke Baker from Gaithersburg, Maryland, competed in the 1986 International Fair, with her project *Malaria Vectors: Eight New Species Implicated* in the Medicine and Health category. She found her specimens, eight various mosquitos, by working with the Walter Reed Army Institute of Research, and ultimately obtained samples from health agencies of the United Nations.

SAMPLE SIZE AND NUMBER OF TESTS

An important part of any experiment is the sample used. Common sense tells us that a procedure done *once* on only *one* subject will barely suggest a trend, much less prove a hypothesis. It is therefore important to have a large enough group to do adequate testing and provide conclusive information. Your results will be more credible if they happened to five subjects than to two, and will be just about conclusive if your sample size was ten. At the same time, you'll want to strike a balance and limit the number of elements to those you can adequately manage.

Another reason for having an adequate sample size is to include a number of extras, "just in case." In case what? In case the plant gets knocked over, the mouse escapes,

someone drinks the soda, or the baby used "it" to make mud pies, that's what!

Obviously, when dealing with live specimens, a larger sample size is more important, because it may be almost impossible to determine whether the individuals in the group were average or abnormal. If you're dealing with seedlings, start with at least 50, and preferably more, to account for the fact that many will not germinate, or may later die.

Certainly, when dealing with human responses to a survey or questionnaire, you must include enough subjects to insure that your results are representative. For example, in Kimberly Lea Quinlan's project on *Teenage Stress* her results were based on 378 responses to a questionnaire. For the project *The Influence of Favorite Color, Sex, Age, Eye and Hair Color Upon Wardrobe Color,* Lan Do compiled the results from 419 surveys.

To make your results at all believable, you'll want to use at least 100 surveys. Incidentally, it's probably a good idea to distribute more than you need, to compensate for those who throw away your survey. Review the tables that follow as a guideline on sample sizes.

Regardless of the number of units, perform a sufficient number of tests or trials to make your results more certain. Statistically, no conclusive evidence can be inferred based on too few trials. Many teachers agree that a minimum of five trials is needed to prove a hypothesis, and even five is a bare minimum, which would be effective only if all the trials yielded the same result. To allow for the "worst possible case," ten trials would be a good minimum.

DOING THE EXPERIMENT

By this time, you probably have all your equipment, and can't wait to begin. Before you start, however, let's take a moment to reemphasize why you're doing this and what you hope to gain.

To achieve your scientific goals and objectives, you must know how to conduct an experiment. Consistency and regularity of procedures and measurements will insure that

Table 6-5. Number of Trials.

Type of Project	Minimum	Suggested
Physics	20	50-100
Animal Behavior	10	25-50
Other	5	20-50

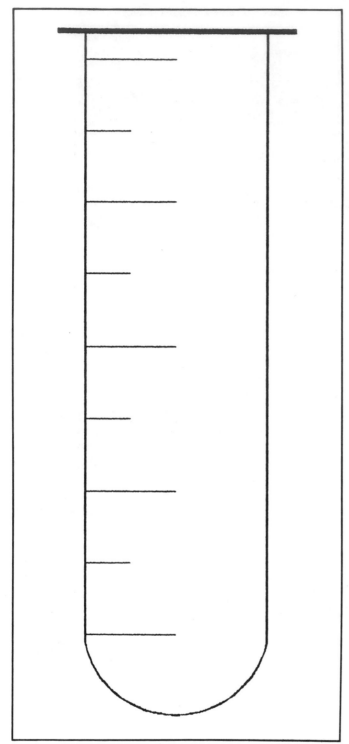

Fig. 6-10. Illustration of test tube.

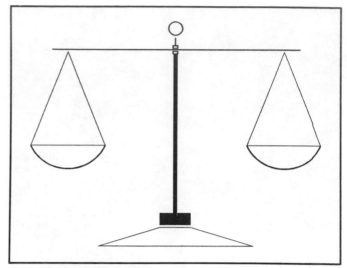

Fig. 6-11. Illustration of scales.

observation and computation is absolutely essential. First of all, show your professionalism by always using the metric system of measurement. Next, be certain that your measurements are exact. Finally, be sure that any instruments are precisely balanced or calibrated.

Wherever possible, you'll want to make a *quantitative analysis* of your results, based on true measurements, rather than simply a *qualitative analysis*, which relies on observation.

For example, "the mice kept in the dark ate twice the amount of food as the ones in the lighted room" is considered a qualitative analysis. In Gwen S. Pearce's project *Stop—You May be Using the Wrong Detergent,* the results

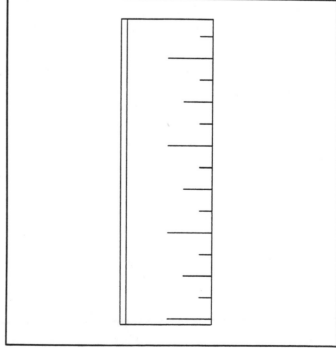

Fig. 6-12. Illustration of ruler.

your data is correct. Understanding and using the experimental scientific method will increase your credibility with teachers and judges, giving them faith in what you've done, regardless of whether your experiment has "worked."

OBSERVATION AND MEASUREMENT

While you are doing your experiment, precise

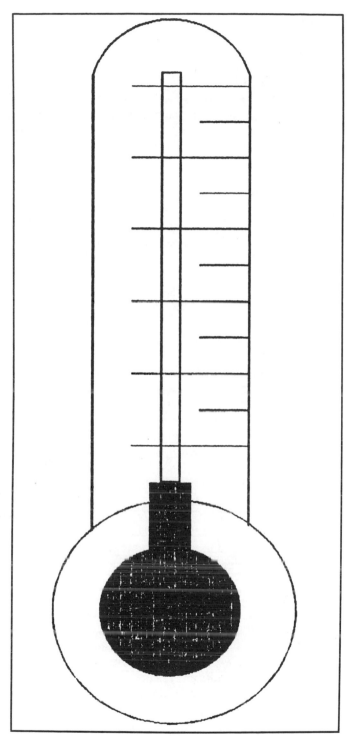

Fig. 6-13. Illustration of thermometer.

Testing Two Theories of Mouse Navigation she used the statistical T-test, thus insuring a quantitative analysis.

If you are using scales, rulers, vessels, or other items that measure weight, size, or volume, choose those that can show the smallest gradations you may require. "A little less than a milliliter," is not only vague, but cannot be compared to a quantity that is "almost a milliliter."

Often, the ability to measure the smallest gradations is essential to the project. For example John Chin-Hung Leu had to make minute, careful measurements of the amphibian eggs under various conditions in his experiment on the *Development of Amphibian Eggs under Simulated Hypogravity Conditions.* Therefore, use tools that will insure the greatest degree of accuracy and also will allow you to make exact comparisons.

RECORD-KEEPING

At the time that you make each and every observation, record your results in writing. This may well be the most important thing you'll do while conducting your tests or trials. A few minutes later, you might forget *exactly* what you saw or what calculations you made. Even if you have a photographic memory, your experiment will lose credibility if your recordkeeping is not current and accurate.

Logs and tables will help keep track of everything that happens during your project. Your tables will help you record the raw data you gather during your experiment. To make it as easy as possible, design a table before you even begin your experiment. Then, you can concentrate on carrying out your procedures and measuring your results without having to worry about "Where do I write that number?" or "One of the mice escaped. Where can I make a note of that?"

When designing your table form, include a place to record the date and time of each entry, as well as whatever measurements and observations you make. Also, leave a final column for notes or comments.

In Table 6-6, you could leave the ML (milliliter) designation out of each column, provided you noted somewhere on the page that all liquid volumes were expressed in milliliters. Note also that a comment such as "Found a hole in container B" would require a more detailed explanation in your diary or log. Such a condition might very well make your observations worthless, especially if you are measuring the amount of liquid that "disappears" in the course of your tests. When dealing with a control group and one or more experimental groups, it's a good idea to have a separate table for each group. This will help you to accurately record the information for each group.

While you're proceeding, don't forget to maintain your time log. This is extremely useful for whatever you do that can't be recorded on your table, for example, locating a graphing program, or conversations you've had with teachers or advisors. Include any additional research material you've read, or perhaps new discoveries you've made

of the cleaning tests with the various detergents were visually observed and evaluated, resulting in a qualitative analysis.

"Group A, kept in the dark room, ate between 8 and 10 cc per day, and group B, kept in the lighted room, ate between 4 and 6 cc per day," is an example of quantitative analysis, is based on exact measurement and is far more accurate. In Ann Wood's project, *Interaction vs. Backup:*

Table 6-6. Sample Experimental Log.

Date	Time	Liq. Absorbed Group A	Liq. Absorbed Group B	Comments
1/27	5:53 P.M.	5ML	5ML	Noticed a hole in container B

Table 6-7. Sample Experimental Log.

Date	Food Remaining (CC's)		
	Group A	Group B	Group C
12/5	6.9	4.8	4.8
12/6	7.1	5.3	3.6
12/7	7.3	5.0	3.9
12/8	7.0	4.8	4.3
12/9	6.8	4.7	4.1

while doing your project. Sometimes, in the course of an experiment, unforeseen problems may develop that cause you to change your procedures. Document any alterations, either in descriptive form, or using photographs, illustrations, sketches, charts, or graphs.

IF ALL ELSE FAILS . . .

For most of you, your experiment is going smoothly. You have measurable observations, which show at least some effects of your testing. If, however, things are going poorly, consider changing your procedures or altering your hypothesis. Jessica Wertlieb overhauled the experimental procedures seven times in the course of her project *Effects of UV Light on F'Plasmid of Escherichia coli,* when the microorganisms did not grow properly. If you started early enough and budgeted your time, you may even be able to change your topic if that becomes necessary.

If you're thinking of taking any of these steps, *please discuss it with your teacher or advisor.* First of all, she or he may realize that if you "hang in there" a little longer, the experiment might still yield some useful data. On the other hand, the teacher may see that by "not getting any results," you are, in fact, concretely disproving your hypothesis, which is equally valuable.

However, if your worst fears are realized and you cannot salvage your experiment, your teacher may be able to help you shift gears with a minimum of extra work. He or she will probably be able to guide you towards an alternate project, which can utilize some of your background research as well as the materials you've accumulated. In any event, you don't need to go it alone! If you've done your work and tried your best (your research, tables and logs will give ample evidence of that), there's always plenty of help available to you.

At this point, it is sometimes tempting to "fudge" your observations or measurements, especially if it appears that your experiment is completely disproving your hypothesis, or, worse yet, doing nothing at all. First of all, let me sympathize. Stuart Allen's experiment on dental adhesives "didn't work," and he was extremely discouraged to find that after all the research and preparation, absolutely nothing happened. He felt that, having "guessed wrong" the project was unsuccessful and he would receive a poor grade. However, because the project was well executed, it not only received a good grade, but he learned some important lessons about science.

Remember that "nothing happening" *is* a result. Rather than revealing the expected effect of the variable on the subject, your experiment may show that your subject is unaltered or minimally changed by the variable in the quantities you used. Although this was not the hypothesis you originally tried to prove, your experiment is equally valid, especially if you followed scientifically sound procedures and kept accurate records throughout. The tests may have refuted the hypothesis, which is as important as proving it.

Rather than disproving your theory entirely, the experiment may prove to be inconclusive. This usually shows that the variable was applied in the wrong quantity, strength, or temperature, or that more tests may be required. Look on the bright side. This gives you a built-in science project topic for next year!

Throughout scientific history, and the history of science projects, additional experimentation and research have turned failed projects into successful ones. Even experiments that work are often refined and enhanced to yield even more significant results.

Robin Rowe of Cocoa, FL, concerned about the prob-

Name_____		
Project Title_____		
		Time Log

Date	Time	Activity
1/25/85	3:35 P.M.	One litter of 8 born in cage B. Removed mother and babies to a holding cage.
2/05/85	4:30 P.M.	Experimented with Chart Master program to graph results.
2/17/85	2:00 P.M. thru	Borrowed plotter and 6 different colored pens to draw graphs.

Table 6-8. Time Log.

lems of water pollution in rural lakes, studied human and cow waste material that contaminated the water. Although she did not fully achieve her project goals, she entered the 36th International Science Fair (in itself, an accomplishment), and won a Third Award. The following year, Ms. Rowe applied the information she had learned about how to actually differentiate between the pollutants. Her project *Differentiating Between Human and Cow Escherichia Coli* was entered in the 37th ISEF. She not only repeated her Third Place Award, but received a four year scholarship to Ohio State University.

As we mentioned earlier, more than the hypothesis or the results and conclusions, it is the experimental design and procedures, including meticulous observation and precise recordkeeping, which are the hallmark of a well-executed project.

Chapter 7

Results and Conclusions

Once the experiment is finished, you'll need to examine and organize your findings. Then, you'll interpret and analyze this data in order to formulate your conclusions.

RESULTS

Results, in terms of a science project, are defined as simply what happened during the course of an experiment. During this phase, the task is not to interpret the results. This will wait until you're ready to work on the conclusion, the true final phase of the experiment. What you'll want to do here is gather, organize, and reorganize the material in various ways, to make the results as clear and meaningful as possible.

The results phase of the project incorporates three elements, the raw data, the smooth data, and the analyzed data.

RAW DATA

The basis of the results will be the raw data, which consists of the tables you developed and maintained while conducting the experiment. It's really tempting to try to combine and analyze the information while you're actually doing the tests. However, it's vital that you simply record your observations. Anticipating the outcome may cause you to unconsciously bias your observations, which would, of course, invalidate the entire experiment.

You may wish to graph your raw results by creating bar and line graphs for your variable and control group(s), extending over the life of the experiment. For the project that we've been following, the effects of sound on mouse appetite, both the actual amount of food eaten and the percentage of available food eaten is shown in both tables and

bar graphs, as illustrated below in Fig. 7-1 and Tables 7-1 and 7-2.

In Mike Iritz's project *The Shaking Earth, The Burning Sky*, the incidence of high magnitude quakes, for the period between 1932 and 1974, was correlated to solar activity. Each factor was graphed separately, for a six year period as shown in Figs. 7-3 through 7-5.

SMOOTH DATA

Smooth data consists of the raw data that has been combined and correlated. For example, if your project examined one control and two experimental groups, whose measurements were charted and graphed daily, you will need to include the data over the life of the project into your smooth data. Figure 7-6 graphs the percentage of food eaten, as shown in Table 7-2 and Fig. 7-2, on a weekly basis.

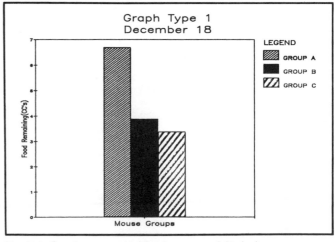

Fig. 7-1. Graph—amount of food consumed (1 day).

Table 7-1. Amount of Food Consumed.

Food Remaining (CC's)	Group A	Group B	Group C
Nov. 20	7.3	3.4	4.7
Nov. 21	6.2	4.3	3.9
Nov. 22	6.8	4.9	3.8
Nov. 23	6.6	5.0	3.7
Nov. 24	6.9	4.3	3.9
Nov. 25	7.1	4.6	4.0
Nov. 26	6.8	4.5	3.9
Nov. 27	6.6	4.1	3.8
Nov. 28	6.7	4.4	4.2
Nov. 29	6.4	4.7	4.1
Nov. 30	6.3	4.6	4.5
Dec. 1	6.5	4.9	4.3
Dec. 2	6.5	5.1	4.1
Dec. 3	6.7	4.7	4.2
Dec. 4	6.6	4.4	4.7
Dec. 5	6.9	4.8	4.8
Dec. 6	7.1	5.3	3.6
Dec. 7	7.3	5.0	3.9
Dec. 8	7.0	4.8	4.3
Dec. 9	6.8	4.7	4.1
Dec. 10	6.2	4.4	3.8
Dec. 11	6.5	4.3	3.5
Dec. 12	6.3	4.6	3.8
Dec. 13	6.7	4.8	4.3
Dec. 14	7.6	4.5	4.6
Dec. 15	7.2	4.1	4.0
Dec. 16	5.8	3.8	3.9
Dec. 17	6.4	3.4	3.8
Dec. 18	6.7	3.9	3.4
Dec. 19	6.9	4.6	3.6
Dec. 20	6.1	4.8	3.9
Dec. 21	6.1	4.7	3.3
Dec. 22	6.4	4.3	3.7
Dec. 23	6.6	4.2	4.2
Dec. 24	7.1	3.7	4.6
Dec. 25	7.4	3.9	4.9
Dec. 26	6.9	4.1	4.0
Dec. 27	6.4	4.3	3.6
Dec. 28	6.0	4.7	3.5
Dec. 29	5.0	5.1	3.2
Dec. 30	6.4	5.3	3.4
Dec. 31	6.7	4.7	3.8
Jan. 1	6.8	4.5	4.1
Jan. 2	6.4	4.0	4.3
Jan. 3	6.5	5.2	4.6
Jan. 4	6.3	4.5	4.0
Jan. 5	6.6	4.3	3.6
Jan. 6	6.9	4.0	3.7
Jan. 7	7.1	3.4	4.3
Jan. 8	7.3	3.7	4.7
Jan. 9	6.8	4.3	4.5
Jan. 10	6.6	4.6	4.2
Jan. 11	6.3	4.8	3.6
Jan. 12	6.5	4.2	3.5

In the earthquake and solar activity correlation project, the data from Figs. 7-3 and 7-4 were combined with the data shown in Fig. 7-5, and on the graphs shown in Figs. 7-7 and 7-8.

As shown in Figs. 7-9, 7-10 and 7-11, the earthquake data for both magnitudes studied, as well as the solar activity, was graphed for the 43 year period studied.

Finally, Figs. 7-12 and 7-13 show the correlation of each magnitude with the solar activity for the entire 43 year period.

If you find that you cannot include all your smooth data on one graph or table, you may use several, provided they combine the experimental and control groups.

Table 7-2. Percentage of Food Consumed.

Percentage of Food Ate	Group A	Group B	Group C
Nov. 20	85.4	93.2	90.6
Nov. 21	87.6	91.4	92.2
Nov. 22	86.4	90.2	92.4
Nov. 23	86.8	90.0	92.6
Nov. 24	86.2	91.4	92.2
Nov. 25	85.8	90.8	92.0
Nov. 26	86.4	91.0	92.2
Nov. 27	86.8	91.8	92.4
Nov. 28	86.6	91.2	91.6
Nov. 29	87.2	90.6	91.8
Nov. 30	87.4	90.8	91.0
Dec. 1	87.0	90.2	91.4
Dec. 2	87.0	89.8	91.8
Dec. 3	86.6	90.6	91.6
Dec. 4	86.8	91.2	90.6
Dec. 5	86.2	90.4	90.4
Dec. 6	85.6	89.4	92.8
Dec. 7	85.4	90.0	92.2
Dec. 8	86.0	90.4	91.4
Dec. 9	86.4	90.6	91.8
Dec. 10	87.6	91.2	92.4
Dec. 11	87.0	91.4	93.0
Dec. 12	87.4	90.8	92.4
Dec. 13	86.6	90.4	91.4
Dec. 14	84.6	91.0	90.8
Dec. 15	85.6	91.8	92.0
Dec. 16	88.4	92.4	92.2
Dec. 17	87.2	93.2	92.4
Dec. 18	86.6	92.2	90.2
Dec. 19	86.2	90.8	92.8
Dec. 20	87.8	90.2	92.2
Dec. 21	87.8	90.6	93.4
Dec. 22	87.2	91.4	92.6
Dec. 23	86.8	91.2	91.6
Dec. 24	85.8	92.6	90.8
Dec. 25	85.2	92.2	90.2
Dec. 26	84.2	91.8	92.0
Dec. 27	87.2	91.4	92.8
Dec. 28	88.0	90.6	93.0
Dec. 29	88.2	89.8	93.8
Dec. 30	87.2	89.4	93.2
Dec. 31	86.6	90.6	92.4
Jan. 1	86.4	91.0	91.8
Jan. 2	87.2	90.4	91.4
Jan. 3	87.0	89.6	90.8
Jan. 4	87.4	91.0	92.0
Jan. 5	86.8	91.4	92.8
Jan. 6	86.2	92.0	92.4
Jan. 7	85.8	93.2	91.4
Jan. 8	85.4	92.6	90.6
Jan. 9	86.4	91.4	90.8
Jan. 10	86.8	90.8	91.6
Jan. 11	87.4	90.4	92.8
Jan. 12	87.0	91.6	93.0

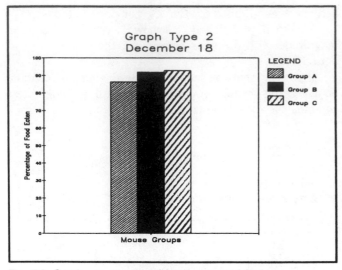

Fig. 7-2. Graph—percentage of food consumed (1 day).

Fig. 7-5. Graph—solar activity—1932-1937.

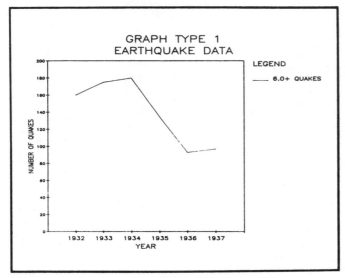

Fig. 7-3. Graph—6.0 earthquake activity—1932-1937.

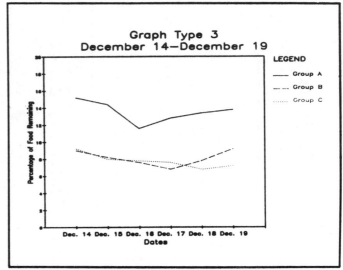

Fig. 7-6. Graph—percentage of food consumed (1 week).

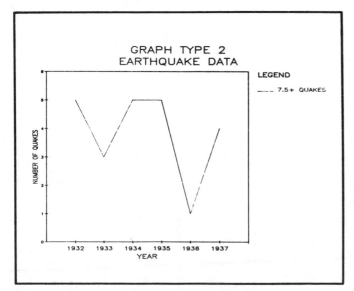

Fig. 7-4. Graph—7.5 earthquake activity—1932-1937.

Fig. 7-7. Graph—combined solar and 6.0 earthquake activity—1932-1937.

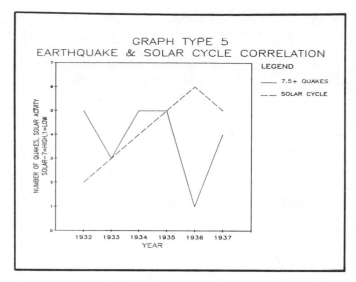

Fig. 7-8. Graph combined solar and 7.5 earthquake activity—1932-1937.

Fig. 7-11. Graph—solar activity—1932-1974.

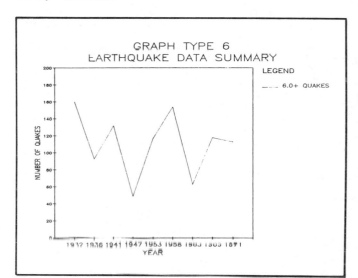

Fig. 7-9. Graph—6.0 earthquake activity—1932-1974.

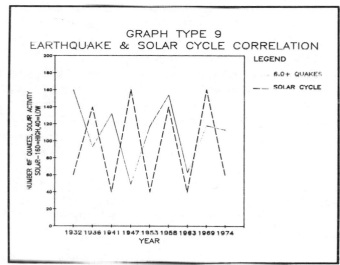

Fig. 7-12. Graph—combined solar and 6.0 earthquake activity—1932-1974.

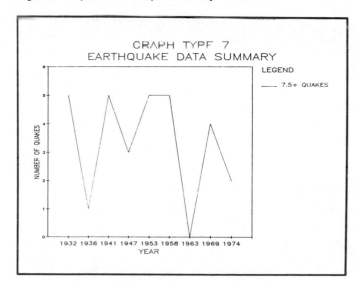

Fig. 7-10. Graph—7.5 earthquake activity—1932-1974.

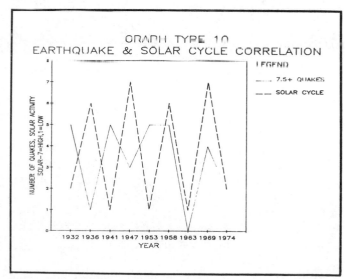

Fig. 7-13. Graph—combined solar and 7.5 earthquake activity—1932-1974.

Computer Note:

Many data base and word processing programs allow you to combine several files into one, making it easier to create smooth or combined data. Be sure, however, to have backup versions of your raw data before beginning work on the smooth data.

The smooth data should also include all the averages, totals, percentages, or other calculations that are necessary to tabulate and correlate your results. Depending on the nature of your experiment, you might require some more involved statistical or mathematical formulas.

If you haven't taken many math or science courses, you may need to consult with your math teacher to help select formulas that will appropriately analyze your raw data. When working with statistics, he or she may also provide guidance to determine how much variance, or which mathematical results, will prove or disprove your hypothesis.

In trying to correlate solar activity with earthquakes, it was insufficient to simply look at the graphs. A statistical test was necessary and the Correlation Coefficient Statistics Test, was used. Based on the data in Table 7-3, the statistical correlation was calculated, as shown in Tables 7-4 and 7-5.

Computer Note:

Your computer can also be very useful to you here. If there's a spreadsheet program available, you can use it to either calculate the results of an equation you've developed or to use one of its own built-in functions, such as count, sum, average, minimum, maximum, or standard deviations. If you've studied programming, you can also write your own programs to compute the statistical results.

Smooth data can also be graphed, by drawing a separate graph for a specified time period while combining the data for the variables and controls, as shown in Figs. 7-3 and 7-4. This, however, will result in a great many graphs if the experiment extended over a long period of time.

At the bottom of each smooth data table or graph, write one or two short paragraphs that summarize the data and explain briefly what the facts and numbers show.

ANALYZED DATA

The last element, analyzed data, will actually show the results, comparing the experimental groups with the control groups. The most effective method of presenting the analyzed data in a clear, accurate, and visually appealing way is to use charts and graphs. Line and bar graphs are the most common types of graphs used to present science project results as shown in Fig. 7-14.

Table 7-3. Raw Data— Earthquakes (6.0 and 7.5) and Solar Activity.

Year	6.0 + Quakes	7.5 + Quakes	Solar Activity 1 = LOW, 7 = HIGH
1932	160	5	2
1933	175	3	3
1934	180	5	4
1935	135	5	5
1936	93	1	6
1937	97	4	5
1938	105	8	4
1939	96	5	3
1940	124	3	2
1941	132	5	1
1942	64	6	2
1943	93	11	3
1944	84	5	4
1945	64	2	5
1946	85	9	6
1947	49	3	7
1948	48	4	6
1949	94	6	5
1950	145	8	4
1951	125	2	3
1952	146	5	2
1953	117	5	1
1954	136	0	2
1955	132	0	3
1956	152	4	4
1957	183	11	5
1958	154	5	6
1959	148	3	5
1960	181	2	4
1961	156	3	3
1962	72	0	2
1963	63	0	1
1964	107	1	2
1965	167	3	3
1966	87	2	4
1967	104	1	5
1968	90	3	6
1969	118	4	7
1970	138	5	6
1971	133	7	5
1972	125	3	4
1973	108	7	3
1974	113	2	2

Regardless of whether you've used graphs in your raw and smooth data sections, here is where they can be used to their best advantage to show the continuity of the results. The objective is to combine as much as possible on one graph to show the correlation of data without a loss of clarity. Consider the use of color, if possible, to distinguish among the various elements. If not, design can be used, as shown in Fig. 7-15.

Pie charts, a very attractive way of diagramming results, will only lend themselves to projects that measure the division of a predetermined and specified quantity, for example, showing the percentage of a group of specimens. If your experiment uses that type of data, pie charts are appropriate and effective.

If you like working with graphs and your data lends itself to this method of presenting results, you might combine several graphing techniques. Remember, however, that as attractive as they are, graphs can only enhance the results; they will not compensate for deficiencies in research or experimentation.

Computer Note:

There are many computer graphics software packages on the market. Some can work on a dot-matrix printer as well as on an elaborate 12-pen plotter. They're a boon for anyone who wishes to include graphs and lacks either the talent or the patience to draw them accurately. Again, investigate using the school's computer if you have a great deal of graphing to do.

If you've used a spreadsheet to perform some of your computations, check to see if the software includes its own graphics package. If so, it will save you the work of recopying your results in order to graph them.

Be sure to title each graph or chart. Describe the data being analyzed and compared. Clearly and accurately label each axis, column, or row, including the unit of measurement used. For illustration, review the graphs shown earlier in this chapter.

The results section can also include any photographs you've taken to show what happened during your experiment. This can be very effective with experiments that have been conducted over several weeks, whose subject is easily and clearly photographed. For example, a project using plants, where the changes occur slowly over a period

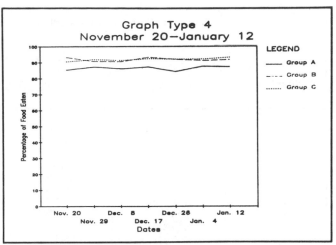

Fig. 7-14. Graph—*Mouse Appetite and Sound*—analyzed data.

of time, would be enhanced by using a series of photos.

Although sketches can add a great deal to the appearance of your project notebook, they are not considered as factual as a photograph. However, if you're talented in that area, by all means, add drawings to liven up your project when you start working on the final notebook or the project display.

Finally, in addition to charts, tables, graphs, and illustrations, you'll have to provide a written description of the results of the experiment. This should contain a narrative summary of observations and measurements. Describe any statistical work you've done, including any correlations between the experimental and control groups. Don't forge ahead and draw your conclusions, give "just the facts." In other words, this short statement should clearly and simply explain what happened and what the results showed.

CONCLUSIONS

At this point, you have done your experiment and organized your results. As you get ready to formulate the

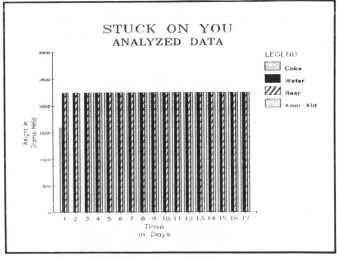

Fig. 7-15. Graph—*Stuck on You* (dental adhesives)—analyzed data.

Table 7-4. Correlating 6.0 Earthquakes and Solar Activity Using Correlation Coefficient.

	Y	X		

6.0 + Quakes vs. Solar Activity

Year	X_1	Y_1	X_1^2	Y_1^2	X_1Y_1
1932	− 1.837	41.907	3.374	1756.167	− 076.983
1933	− 0.837	56.907	0.701	3238.407	− 047.631
1934	0.163	61.907	0.026	3832.477	0.10.091
1935	1.163	16.907	1.352	0285.847	019.663
1936	2.263	− 25.093	4.678	0629.658	− 054.276
1937	1.163	− 21.093	1.352	0444.915	− 024.531
1938	0.163	− 13.093	0.026	0171.426	− 002.134
1939	− 0.837	− 22.093	0.701	0488.101	018.492
1940	− 1.837	05.907	3.374	0034.892	− 010.851
1941	− 2.837	13.907	8.048	0193.404	− 039.454
1942	− 1.837	− 54.093	3.374	2926.053	099.368
1943	− 0.837	− 25.093	0.701	0629.658	021.003
1944	0.163	− 34.093	0.026	1162.332	− 005.560
1945	1.163	− 54.093	1.352	2926.053	− 062.901
1946	2.163	− 33.093	4.678	1095.146	− 071.580
1947	3.163	− 69.093	10.004	4773.843	− 218.541
1948	2.163	− 70.093	4.678	4913.028	− 151.611
1949	1.163	− 24.093	1.352	0580.472	− 028.020
1950	0.163	26.907	0.026	0723.986	004.386
1951	− 0.837	06.907	0.701	0047.706	− 005.781
1952	− 1.837	27.907	3.374	0778.801	− 051.265
1953	− 2.837	− 01.093	8.048	0001.195	003.101
1954	− 1.837	17.907	3.374	0320.661	− 032.895
1955	− 0.837	13.907	0.701	0193.404	− 011.640
1956	0.163	33.907	0.026	1149.684	005.526
1957	1.163	64.907	1.352	4212.918	075.486
1958	2.163	35.907	4.678	1289.313	077.667
1959	1.163	29.907	1.352	0894.428	034.781
1960	0.163	62.907	0.026	3957.291	010.253
1961	− 0.837	37.907	0.701	1436.941	− 031.728
1962	− 1.837	− 46.903	3.374	2124.564	084.672
1963	− 2.837	− 55.903	8.048	3035.238	156.298
1964	− 1.837	− 11.093	3.374	0123.054	020.377
1965	− 0.837	48.907	0.701	2391.894	− 040.935
1966	0.163	− 31.093	0.026	0966.774	− 005.068
1967	1.163	− 14.093	1.352	0198.613	− 016.390
1968	2.163	− 28.093	4.678	0789.216	− 060.765
1969	3.163	− 00.093	10.004	0000.008	− 000.294
1970	2.163	19.907	4.678	0396.288	043.058
1971	1.163	14.907	1.352	0222.218	017.336
1972	0.163	06.907	0.026	0047.706	001.125
1973	− 0.837	− 10.093	0.701	0101.868	008.447
1974	− 1.837	− 05.093	3.374	0025.939	009.355

E = SIGMA = the sum of . . .

$$0.143 = \frac{\mathrm{E}x_1Y_1}{\sqrt{(\mathrm{E}x_1^2)\,(\mathrm{E}y_1^2)}}$$

"No Correlation, Statistically"

conclusions, you may wonder, "how do conclusions differ from results?"

Your results, as we've just discussed, stated what happened in your experiment, including any necessary mathematical or statistical interpretations and data correlation. You've also created charts, tables, and graphs to illustrate the data, and make it easy to see and interpret.

The word "interpret" brings us to the last scientific step in your project; the conclusions. Since the beginning of your project, as you researched the background material, and executed your experiment, the purpose was to answer a question and to prove the hypothesis. When you draw your conclusions, you will be analyzing your results to determine what you've accomplished, in other words, what was really learned from the trials and testing (or trials and tribulations, if you prefer!) of your experiment.

Year	X_1	Y_1	$X_1{}^2$	$Y_1{}^2$	X_1Y_1
1932	− 1.837	0.907	3.374	00.823	− 01.666
1933	− 0.837	− 1.093	0.701	01.194	00.915
1934	0.163	0.907	0.026	00.823	00.148
1935	1.163	0.907	1.352	00.823	01.055
1936	2.263	− 3.093	4.678	09.576	− 06.690
1937	1.163	0.093	1.352	00.008	− 00.108
1938	0.163	− 3.907	0.026	15.265	00.637
1939	− 0.837	0.907	0.701	00.823	− 00.759
1940	− 1.837	− 1.093	3.374	01.194	02.008
1941	− 2.837	0.907	8.048	00.823	− 02.573
1942	− 1.837	1.907	3.374	03.637	− 03.503
1943	− 0.837	6.907	0.701	47.707	− 05.781
1944	0.163	0.907	0.026	00.823	00.147
1945	1.163	− 2.903	1.352	04.381	− 02.434
1946	2.163	4.907	4.678	24.079	10.614
1947	3.163	− 1.093	10.004	01.194	− 03.457
1948	2.163	− 0.903	4.678	00.008	− 00.201
1949	1.163	1.907	1.352	03.637	02.217
1950	0.163	3.907	0.026	15.265	00.636
1951	− 0.837	− 2.093	0.701	04.381	01.752
1952	− 1.837	0.907	3.374	00.823	− 01.666
1953	− 2.837	0.907	8.048	00.823	− 02.573
1954	− 1.837	− 4.093	3.374	16.753	07.519
1955	− 0.837	− 4.093	0.701	16.573	03.425
1956	0.163	− 0.093	0.026	00.008	− 00.015
1957	1.163	6.907	1.352	47.707	08.033
1958	2.163	0.907	4.678	00.823	01.962
1959	1.163	− 1.093	1.352	01.194	− 01.271
1960	0.163	− 2.093	0.026	04.781	− 00.341
1961	− 0.837	− 1.093	0.701	01.194	00.915
1962	− 1.837	− 4.093	3.374	16.753	07.519
1963	− 2.837	− 4.093	8.048	16.753	11.612
1964	− 1.837	− 3.093	3.374	09.576	05.682
1965	− 0.837	− 1.093	0.701	01.194	00.915
1966	0.163	− 2.093	0.026	04.381	− 00.341
1967	1.163	− 3.093	1.352	09.576	− 03.597
1968	2.163	− 1.093	4.678	01.194	− 02.064
1969	3.163	− 0.093	10.004	00.008	− 00.294
1970	2.163	0.097	4.678	00.823	01.962
1971	1.163	2.907	1.352	08.451	03.381
1972	0.163	− 1.093	0.026	01.194	− 00.178
1973	− 0.837	2.907	0.701	08.451	− 02.433
1974	− 1.837	− 2.093	3.374	04.381	03.845

Table 7-5. Correlating 7.5 Earthquakes and Solar Activity Using Correlation Coefficient.

$$\begin{array}{cc} Y & X \end{array}$$
7.5 + Quakes vs. Solar Activity

E = SIGMA = the sum of. . .

$$0.182 = \frac{Ex_1Y_1}{\sqrt{(Ex_1{}^2)\, Ey_1{}^2}}$$

"No Correlation, Statistically"

The project conclusions should be stated and explained in a paper of approximately three pages. In it, you'll analyze the results in light of your original assumptions, giving an interpretation of the data. You'll also critique your own project techniques, including the experimental design itself, and suggest future improvements on it.

Begin your conclusion paper by restating your question or hypothesis. Next, compare the results to your original hypothesis. The result data may concretely establish your theory to be true or false. However, it is also possible that the results were inconclusive, which means that although there may be a trend in your data, it is not strong enough to prove or disprove your hypothesis.

In formulating conclusions, patterns are what you're looking for. Closely examine your tables, graphs, and charts to see if a trend clearly emerges. The most important thing is to review your results critically and *without bias* in order to reach a definitive conclusion.

Fig. 7-16. Terri Newport; *Rehabilitation of Ground Water Aquifers.*

If the results show a conclusive direction, you can happily state either that your assumption was correct and you've proven your hypothesis, or that the experiment failed to demonstrate that your theory was true. In any event, explain why your results occurred. This is important even if the hypothesis was correct.

For the project correlating mouse appetite and sound, the experiment proved the hypothesis, as shown by the consistently greater amounts and percentages of food eaten by the subjects in the experimental groups. (Refer to Fig. 7-14.)

You'll also want to explain results that disprove the hypothesis. The most likely reason is that the theory was incorrect. That's fine—a hypothesis is simply an educated guess, and refuting it is as scientifically valid as conclusive proof. In the earthquake and solar correlation project, although the graphs seemed to indicate a relationship, as shown in Figs. 7-7 and 7-8, the Correlation Coefficient Statistical Test did not. The statistical test required $+ .82$ or $- .82$ to indicate a relationship. Because the calculated result, as shown in Table 7-4, was 0.143, and in Table 7-5, the calculated result was 0.182, there was no statistical correlation at all.

Another possible reason for inconclusive results is a weakness in the experimental design itself. In the course of a science project lasting several weeks, which is conducted in a home environment rather than a fully equipped lab, it is sometimes impractical to perform a sufficient number of tests on enough specimens to overwhelmingly prove something.

The experiment graphed in Fig. 7-15 hypothesized that an orthodontic adhesive would weaken in various substances that were high in sugar. With one exception, the bonds did not break under pressure. The participant concluded that although the theory was disproved, it may have been because the trials were not carried on long enough, or the pull to loosen the bond was too weak.

In writing your conclusions, use your research as well as your results to explain the conclusions reached. The clearest and easiest way to do this is to discuss each fact or occurrence in a separate paragraph, referring to the experimentation and analysis that was done to reach the conclusion. Finally, write a summary that restates the hypothesis and conclusion as supported by the results.

Fig. 7-17. Caroline K. Horton; *Fly Ash Foam: A Method to Create a Ceramic Material from the Waste Products of Coal Incineration through the Use of Microwaves.*

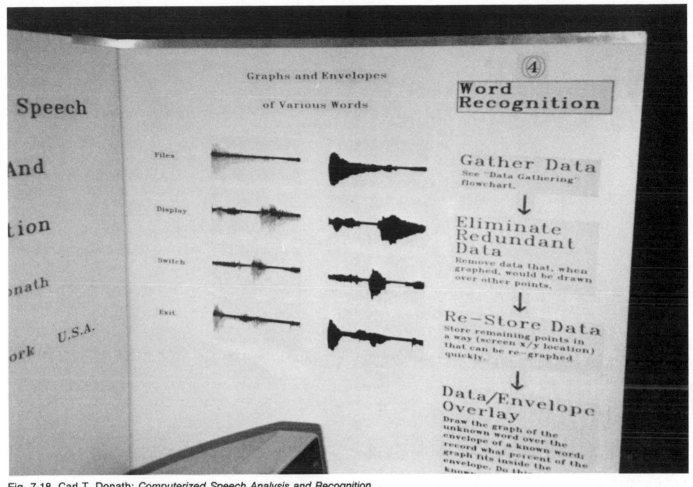

Fig. 7-18. Carl T. Donath; *Computerized Speech Analysis and Recognition.*

IMPROVEMENTS AND ENHANCEMENTS

Regardless of whether the project achieved its objectives, you'll need to analyze your experiment. *Realistically* reexamine the steps of your procedure in light of your results. Discuss the strengths and weaknesses of the design. Be honest—there are always some of each! Finally, suggest future modifications or improvements.

If this project has inspired you to continue research and experimentation in this field, briefly state how you plan to proceed. Even if you do not intend to pursue the subject, perhaps because it is beyond your resources or abilities, speculate on further work that could be done. However, keep in mind that several of the big winners at ISEF had been working on variations of the same project for many years, refining and developing their work, building each year on the strengths and weaknesses of their prior year's project.

Terri Newport of Stratford OK entered the 36th ISEF with a project entitled *Ground Water Pollution—Identification, Isolation and Rehabilitation*. The project did not win any awards, but the following year, she worked in the same general category. She expanded and enhanced the scope of her project and built actual models for her 37th ISEF entry *Rehabilitation of Contaminated Ground-Water Aquifers: Methods, Considerations and Constraints*. This project won her six awards, including the coveted four year scholarship to Ohio State University

"REAL LIFE" APPLICATIONS

A powerful ending to your conclusions paper is a discussion of any potential practical value that your experiment may have. This will show teachers and judges not only your analytical ability as it relates to scientific reasoning, but a well-rounded approach, relating your work to other fields of endeavor. Looking at those types of relationships have brought a great deal of personal satisfaction to many students. In some instances, there is even the potential for professional recognition and commercial reward.

Caroline K. Horton, whose objective in her engineering project *Fly Ash Foam: A Method to Create a Ceramic Material from the Waste Product of Coal Incineration through the use of Microwaves* produced a product whose qualities exceeded that produced at the Los Alamos National Laboratories, where she was conducting her research. She is now applying for a patent for her procedure.

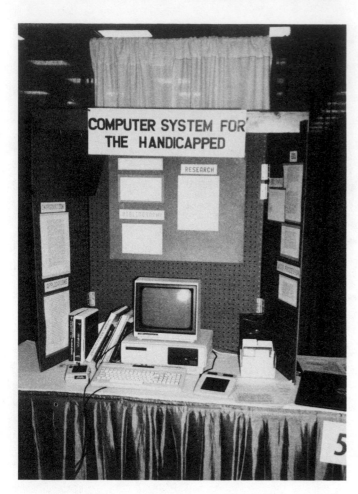

Fig. 7-19. Scott Smith; *Computer System for the Handicapped.*

On the other hand, both Carl Donath, with his *Computerized Speech Analysis and Recognition* and Scott Smith, with his *Computer System for the Handicapped,* have received a less tangible, but very real reward.

They can take particular satisfaction in the possible contribution their projects can make towards using the growing computer technology to make a better life for others.

Chapter 8

Best Foot Forward!

"Clothes don't make the man," says the old expression, but they can certainly help dress him up! Now that you've done the research, completed your experiment, analyzed and graphed the results and formulated your conclusions, you're ready to present your project.

As with any other presentation, it should be complete and accurate, as well as something that will generate immediate interest in your work. Although these might seem like opposing goals, they can both be satisfied, because the project presentation consists of two parts, a science fair notebook, and the display, or backboard.

THE SCIENCE FAIR NOTEBOOK

The Science Fair Notebook will contain all written and visual materials associated with your project. If this project was required, there's some good news and bad news. The bad news is that the notebook will probably constitute the lion's share of your grade, but the good news is that almost all the work is already done. Now, it's simply a matter of getting everything into the most attractive and presentable form.

You will want to put your best foot forward in this notebook, so start out by using the appropriate cover. Your teacher may have his or her own specific requirements for the size and type of folder, but if not, a good guideline is to use 3-hole, 8 1/2 × 11 paper and folder.

Use regular bond paper, since erasable bond smudges, and thin, onion skin paper may cause the reader to see double. *Always* double-space all written work, with the possible exception of tables, bibliography entries, or footnotes written in standard form, as shown in Chapter 4.

Some teachers may allow handwritten notebooks. If so,

use ink only (no erasables please—they smudge!). The only possible exception is the log section, which represents raw data. However, type or use computer printouts wherever possible. The easier your work is on the teacher's eyes, the happier he or she will be. And a happy teacher usually gives better grades!

Now, let's review the items to be included in your science fair notebook.

1. Required Forms

The forms include the entry form, with the project description shown in Fig. 8-1.

If you used live vertebrates or tissue samples in your experiment, include the required forms (refer to Figs. 6-5 and 6-6). If you used human specimens, include the forms, as shown below.

2. Title Page

3. Table of Contents

Remember that although the Table of Contents goes right in the beginning of the notebook, it will probably be the last thing you'll create, once the pages are numbered. An example of a good format is shown in Fig. 8-4.

4. Abstract

The abstract is a shortened, or abbreviated, version of your background research paper. It should be about 200 words, and should summarize the hypothesis, the procedure, and the results. Keep in mind that you may need to fit the abstract, or some shortened version of it, on your entry form. Refer back to the project summary in Fig. 8-1.

GREATER SAN DIEGO SCIENCE AND ENGINEERING FAIR (GSDSEF)

CERTIFICATE OF ENTRANCE

(TYPE OR PRINT CLEARLY)

OFFICIAL
USE
ONLY

NAME OF ENTRANT (LAST, FIRST, MIDDLE)

HOME ADDRESS

GRADE _____ FEMALE ☐ PROJECT ADVISOR (First Initial, Last Name)   CITY, STATE, ZIP

AGE _____ MALE ☐

SCHOOL   CITY   ENTRANT'S SOCIAL SECURITY #   HOME PHONE
                                                (Optional — Helpful to Awards Staff)

HAVE YOU EVER HAD A PROJECT ACCEPTED
BY THE GSDSEF **BEFORE** THIS YEAR?   NO ☐   YES ☐   PARENT(S) / GUARDIAN(S) NAME(S)

IF YES, HOW MANY YEARS? _____

1. Biology / Microbiology	7. Chemistry / Biochemistry
2. Botany	8. Earth / Space Sciences
3. Zoology	9. Physics
4. Medical Sciences	10. Engineering / Electronics
5. Human Psychology / Social Sciences	11. Mathematics
6. Animal Behavior	12. Computers

WAIVER — ALL FAIR ACTIVITIES INCLUDING TOURS
IN CONSIDERATION OF YOUR PERMITTING THE UNDERSIGNED STUDENT TO TAKE PART IN THE GREATER SAN DIEGO SCIENCE AND ENGINEERING FAIR (GSDSEF), WE WAIVE ALL CLAIMS AGAINST THE GSDSEF, UNION / TRIBUNE PUBLISHING CO., THE COPLEY PRESS INC. AND THEIR RESPECTIVE OFFICERS, DIRECTORS, GOVERNORS, COMMITTEES, EMPLOYEES AND AGENTS FOR INJURY TO OR DEATH OF PERSONS OR LOSS OR DAMAGE OF PROPERTY IN ANY WAY OCCURRING IN CONNECTION WITH THE GSDSEF AND WE AGREE TO INDEMNIFY AND HOLD THEM HARMLESS AGAINST ALL SUCH LIABILITY.
WE ALSO AGREE THAT THIS PROJECT WILL BE KEPT ON DISPLAY AT THE FAIR UNTIL 5:00 P.M. SUNDAY AND WILL BE REMOVED BY 6:00 P.M. SUNDAY.

Parent/Guardian's Signature _____

Date _____

DISPLAY AREA REQUESTED (CHECK ONE)   ELECTRICAL OUTLETS REQUIRED
(MAXIMUM SIZE — 122cm x 76cm or 48 in. x 30 in.)   (110V 60 CYCLE ONLY)
☐ TABLE   NUMBER OF OUTLETS _____
☐ FLOOR (NO TABLE)   AMPS _____ WATTS _____

DOES PROJECT INVOLVE EXPERIMENTS WITH LIVE   DOES PROJECT INVOLVE EXPERIMENTS WITH   DOES PROJECT INVOLVE EXPERIMENTS WITH
VERTEBRATE ANIMALS? _____   HUMAN SUBJECTS? _____   TISSUE SAMPLES OR PARTS OF HUMAN
(LIVE VERTEBRATE ANIMALS MAY NOT BE DISPLAYED)   OR VERTEBRATE ANIMALS? _____

NOTE: IF LIVE VERTEBRATE ANIMALS ARE USED,   NOTE: IF HUMAN SUBJECTS ARE USED,   NOTE: IF TISSUE SAMPLES / PARTS ARE USED,
GSDSEF FORM 2 MUST BE SUBMITTED WITH   GSDSEF FORM 3 MUST BE SUBMITTED   GSDSEF FORM 4 MUST BE SUBMITTED
THIS CERTIFICATE   WITH THIS CERTIFICATE   WITH THIS CERTIFICATE

CERTIFICATION OF COMPLETENESS AND ACCURACY OF INFORMATION AND COMPLIANCE WITH ALL REGULATIONS

SIGNATURE OF ENTRANT   SIGNATURE OF SCHOOL PROJECT ADVISOR

PROJECT TITLE (10 OR FEWER WORDS)

PROJECT SUMMARY — (200 - 250 WORDS DESCRIBING PROBLEM, METHODS AND SIGNIFICANT FINDINGS, NEATNESS AND THOROUGHNESS IMPORTANT.)

Sample

TYPE SINGLE SPACE TO DOTTED MARGIN

GSDSEF FORM 5 (1987)

LIST **ALL** PROFESSIONAL ORGANIZATIONS, SCIENTISTS AND OTHERS, (IF PARENTS, NEIGHBORS — NO NAMES PLEASE)
FROM WHOM YOU WERE ABLE TO OBTAIN ADVICE AND / OR HELP IN DOING YOUR PROJECT.

Fig. 8-1. Project entry form.

CERTIFICATION OF COMPLIANCE OF RESEARCH INVOLVING HUMAN SUBJECTS

Name of Entrant _____ School _____

Project Title _____

Because federal regulations have become increasingly more rigid, students must plan carefully before undertaking research which involves the use of human subjects in either behavioral or biomedical studies. This will protect subjects from unnecessary exposure to physical or psychological risks and experimenters and schools from legal complications.

A **human subject** is legally defined as:

A person about whom an investigator (professional or student) conducting scientific research obtains (1) data through intervention or interaction with the person or (2) identifiable private information.

A **subject at risk** is legally defined as:

Any individual who may be exposed to the possibility of injury, including physical, psychological or social injury, as a consequence of participation as a subject in any research...

Students using human subjects must comply with all regulations that reflect the will of society and plan proper methodology for the protection of those subjects. It is essential that they be alert to humane concerns at all times.

The following steps must be taken **before** any student begins research involving human subjects:

1) The student completes the "Research Plan" section of this form (GSDSEF-3) and submits it to the sponsoring teacher.

2) The sponsoring teacher reviews the "Research Plan" and determines if **ANY POTENTIAL** physical, psychological or social risk is involved (as defined in **subject at risk** above).

   a) If none is apparent, the teacher signs the certification. (No additional certification is necessary.)

   b) If any question exists, the student must redesign the experimental study or plan a different study.

   NOTE: If assistance is needed to evaluate a particular research plan, call the appropriate screening committee representative (as listed in the GSDSEF Student Guide.)

NOTE: Any project involving human subjects that is developed with the advice and assistance of personnel at a medical/scientific organization must comply with any regulations of that organization requiring approval of its Institutional Review Board and Informed Consent Certification.

RESEARCH PLAN

Describe proposed experimental procedures:

Explain why human subjects are proposed for this experimentation:

Fig. 8-2. Certification of compliance of research involving human subjects.

Describe and assess any potential risk (Physical, psychological, social, legal or other):

Describe the potential benefits to the individual or society:

Signature of student _____ Date _____

CERTIFICATION

CERTIFICATION BY TEACHER/ADVISOR of compliance with federal regulations for the protection of human subjects in behavioral and biomedical research. (**Must** be completed **before** the start of experimentation.)

I certify that, upon reviewing this research plan, I found that the experimental procedures constitute no physical, social or psychological risk to either experimenter or subjects.

I agree to supervise this experimentation and will insure that it is conducted in a humane, risk-free manner.

Signature _____ Name (type or print) _____ Date _____

Institution _____ Title _____

Institution Address _____ Phone _____

Home Address _____ Home Phone _____

This form, properly completed, must be part of the carefully planned procedures for any experiment involving human subjects. It must accompany any such project exhibited at or presented for any public display associated with the Greater San Diego Science and Engineering Fair.

Form GSDSEF-3 (1987)

HOW MUCH DO WE REALLY KNOW ABOUT SEX?
By:
Erin Filner

Crawford High School
P. Wright
February 28, 1986

Fig. 8-3. Title page.

5. Background Research

The Background Research Paper is the same research paper you finished at the beginning of the project. However, you probably had corrections marked on it by your English and Science teachers. Incorporate their suggestions now, in order to fix, clarify, and edit your work.

You might have done some additional research as you progressed with the experiment, especially if you got some professional advice and assistance. Incorporate this information into the final paper to make it more thorough and complete.

6. Bibliography

Just as with the background research paper, the bibliography should be almost complete. Make any corrections in format, spelling, or style that your teachers have requested. If you've added information to the research paper, or conducted interviews, etc., remember to credit any new or additional sources.

7. Statement of the Problem or Question

Restate the question or problem that expresses the purpose of the experiment.

8. The Hypothesis

What was the educated guess that you made, which explained what you tried to prove with the project? Check to see that you've identified the subject of the experiment that you separated into experimental and control groups. Make certain that you've named the dependent and independent variables, and the controls. If your teacher asked you to reword it when you handed it in the first time, now is the time to make these corrections.

Please be honest and state your *original* hypothesis even if the experiment disproved it. Remember that this information is just as valuable as a correct guess!

9. Procedures

List all your procedures, numbering them sequentially. Avoid the first person, use the past tense and include diagrams, where necessary. If you altered the procedures in any way since the time they were originally approved,

be sure to modify them now, before including the list in your final notebook.

10. Materials List

Include anything and everything you bought, begged, borrowed, or built.

11. Variables and Controls

Fully describe each variable and control. Explain its role in the experiment and show how it was managed and/or monitored.

12. Results

Include raw, smooth, and analyzed data, in all forms, including all the charts, graphs, tables, photographs, and diagrams you've created or collected in the course of the experiment. Be sure that all results, in whatever format, are neat and legible. Accurately and clearly label and title all material. Include the narratives that describe your observations and the outcome of the experiment, as shown in Chapter 7.

13. Conclusions

Since you've just finished the conclusion paper, all you'll need to do is make any necessary corrections and type or print the final copy.

14. Acknowledgments

Here, you get the opportunity to thank everyone who has helped you to complete your project. It is a good idea not to identify teachers, school, or family by name, because at certain levels of competition, you are required to remain anonymous. See Fig. 8-5 for an example of a well-written, anonymous acknowledgement section.

Fig. 8-4. Table of contents.

ACKNOWLEDGEMENTS

I would like to thank, for helping me with my science project this year, my mother, my father, and my brother.

I would also like to thank   of the San Diego State University Natural Sciences department, for helping me find the information that I needed to do my science project.

I would further like to thank the people at the San Diego branch of the National Oceanographic and Atmospheric Administration at the Southwest Fisheries Center and the librarians at the UCSD Scripps and SDSU libraries for helping me with the search for information within their library.

Last and definitely not least, the people I could not possibly forget for giving me all the support and help I needed: , my science teacher, and   ,my algebra teacher, for helping me with the statistics correlation test.

Fig. 8-5. Acknowledgments.

15. Project Log

The Project Log is one place where the "neatness counts" rule doesn't necessarily apply. Include your diary, working log, rough notes, and drafts of tables and graphs.

When you've reprinted and collected everything, review all your materials at least once, preferably with someone helping you. Make sure that you have all the sections in order, and that the pages within each section are in the proper sequence.

Number the pages and create the table of contents. If you're going to number your pages by hand, have someone help you by writing the numbers on the page, if your handwriting isn't especially legible. Another alternative is to insert each page into the typewriter to put the numbers either on the top or bottom of each page.

Computer Note:

If you have all the information on diskettes, obviously you only need to make corrections and reprint, taking special care to have proper margins and format. At this time, consider using a fresh ribbon in your printer.

Page numbering could also be done using your word processor or other software. Consult the user manuals to learn how to include different files in one document. If your notebook materials come from various sources, be sure to allow for inserts when deciding to let the program number the pages.

Now that you're done with your notebook, let me repeat that a neat, well-presented notebook will certainly enhance a well-done project, remembering that it cannot compensate for a poorly researched, inadequate experiment.

THE SCIENCE FAIR DISPLAY

The final step is to construct your project exhibit. This presentation is not your product, the research and experiment are. What this represents is your advertisement. A good display will attract attention to your work, inviting judges to look at your notebook and investigate your project more thoroughly.

Just as with the notebook, the most attractive, elaborate backboard in the world won't substitute for mediocre work. However, because you've followed this step-by-step guide, your project will be excellent, deserving a first-class exhibit.

The purpose of the display is to summarize your project. Read that sentence again and remember the key word, *summarize*. Do not try to recreate your entire project notebook. Instead, simply cover the main points and the highlights.

Most science project displays consist of three-sectioned, free standing backboards. The sections are normally hinged together, for easier folding and transporting to the various science fairs where you'll be competing. However, backboards with as few as two sections, or as many as four, have been used successfully.

When looking at fairs at all levels, we've seen backboards made of a variety of elements. The best are constructed of rigid, durable, fireproof goods, such as masonite, pegboard, plywood or plexiglas. Some science fairs may prohibit displays made of cardboard or other paper products, due to fire regulations. Although the actual construction may require some time and expense, the basic backboard can later be stripped of the display materials and used for subsequent science projects.

The normal height for a science fair display is between 3 and 5 feet. When deciding, first check to see if the science fair itself has restrictions on overall size. These limits are usually imposed due to the size of tables available. Occasionally, there is a display that is so large that it cannot be placed on a table. This happens most often with engineering projects. Be sure to check with those in charge of your local science fair before designing such a presentation.

Next, decide whether or not you plan to cover the basic backboard or keep it bare. If you're using pegboard, you may wish to use hooks to attach material. Some students

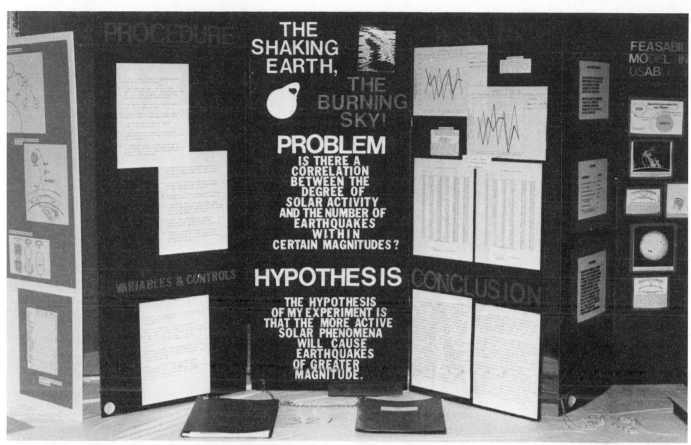

Fig. 8-6. Mike Iritz; *The Shaking Earth, The Burning Sky.*

Fig. 8-7. John Chin-Hung Leu; *The Effects of Microgravity on Embryogenesis.*

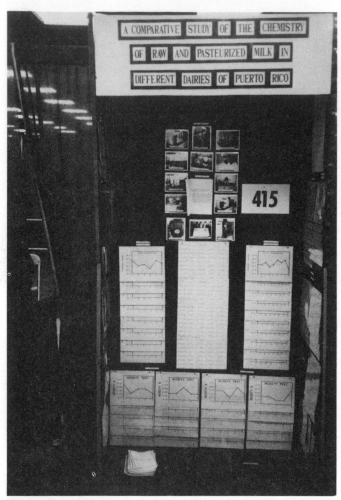

Fig. 8-8. Francis Vazquez; *Comparative Study of Raw and Pasteurized Milk from Different Dairies of Puerto Rico.*

have painted their backboards attractively, and others glued velcro strips to the basic backboard, in order to attach materials later. If you do plan to cover the panels, keep in mind that construction paper comes in 12-inch sheets, which will make it harder to cover an odd-sized backboard. Other covering materials may also come in standard sizes, which should be considered when choosing the size.

Your display *must* include summaries of the problem, hypothesis, procedures, results, and conclusions. You can do this using all written material, or you may include graphs, photographs, drawings, tables, or other artwork. Remember, however, that all this information must fit onto the backboard you have selected. However, some students have very creatively solved the problem of too much material and too little room on a backboard. Francis Vazquez created rows of overlapping flip charts to display his many graphs on the backboard for his project *A Comparative Study on the Chemical and Bacteriological Phases in the Processing of Raw and Pasteurized Milk from some Dairies in Puerto Rico.*

In the display for *Effect of pH Upon the Culture: Penicillum Notatum* Shawn Bray not only created a "mini-diorama" in front of the backboard to display the molds, but

also created a panel to drape over the front of the display table, in which pockets for the Log Book, Abstract, and Research were included.

The best way to insure a well-designed backboard is to make a map or blueprint of your project display. Decide what to put on each of the three panels. Many have included the title, the question and the hypothesis on the center panel, reserving the sides for the procedures, results and conclusions. However, there's no hard and fast rule for this, as you can see from both the photographs and the sample backboard layouts included in Fig. 8-11 A and B.

Be sure to figure out the dimensions of everything to be included, as well as the size and style of the lettering you will use. One thing to keep in mind is that when displaying at a science fair, your notebook will be attached to the lower left portion of the middle section by a piece of chain, twine or other strong material. Be sure to leave a small spot free for that attachment.

This layout is strictly for your use, so feel free to create several versions, until you have one that is suitable and attractive. Once you've decided on a blueprint, show it to your teacher before you actually begin mounting things on the backboard.

Fig. 8-9. Shawn Bray; *Effect of pH upon the Culture—Penicillin Notatum.*

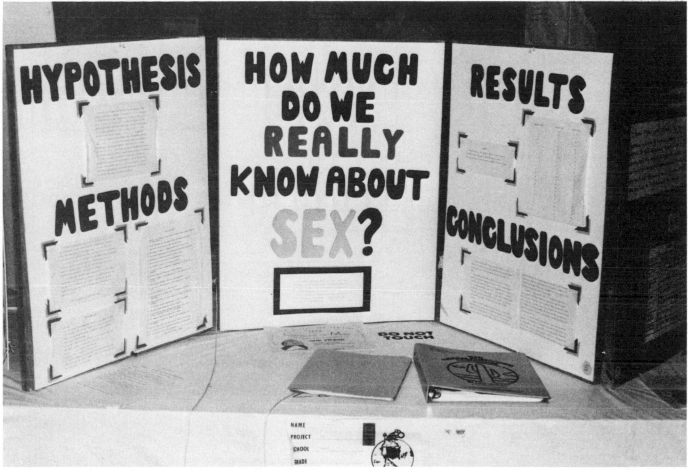

Fig. 8-10. Erin Filner; *How Much Do We Really Know About Sex?*

If you plan to display any equipment or specimens such as plants on the table in front of the backboard, plan to leave the bottom portion clear of text and illustrations, if necessary, because it may be blocked by these items. Notice that David Rodgero, when designing the display for his project *Growth Pattern of Wheat in Relation to Water Source* placed his written and visual material on the upper two-thirds of the backboard, leaving room for four trays of wheat, and tags for each.

In Caleb John's project *What Effect does Drafting have on Energy Requirements,* he stopped the written material about 3/4 of the way down the backboard to allow room to display the equipment.

Of course, if your material can be placed flat upon the display table, as was done in Kelly O'Neill's project *Which Wood Burns Best in a Fireplace,* you need not allocate any extra space for it.

Remember, however, that no live insects or vertebrates can be displayed at a science fair.

As you've noticed, at least some portions of your display will consist of written material. The problem and hypothesis, which are already short statements, can probably be used without any further editing. However, you can show display printed matter in several ways.

—Typewritten.
—Computer Printout. You can experiment with different sizes and fonts using your computer.
—Hand lettering.
—Stencilling or press-on letters.
—Lettering made on a Kroy or other labelling machine.

As shown in the sample display blueprints, be sure to use a title on everything you display, using larger lettering to make it stand out. When doing the written material, you will become an expert in the art of summarizing.

Some participants create backboards which consist mainly of written material, as in *The Study of the Wasp.* However, Akira Soya used illustrations and graphs with the text to create an interesting display.

The main thing to remember here is that "less is more." You'll want to be accurate using as few words as possible.

It's unnecessary to present all the facts on the display. Simply highlight the most important points. If you've done this well, the judges will take the time to examine the notebook and interview you to find the details. On the other hand, including too much information will make it look busy and difficult to read, perhaps discouraging a closer look.

Fig. 8-11. Two styles of backboard layouts.

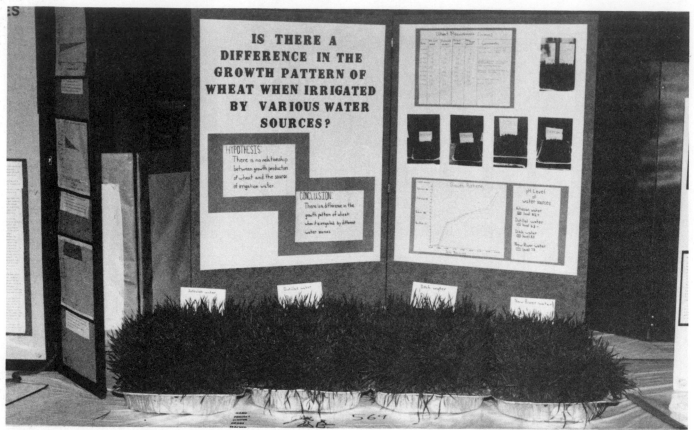

Fig. 8-12. David Rodgers; *Is There A Difference in the Growth Pattern of Wheat When Irrigated by Various Water Sources?*

If you decide to include graphs, charts, or tables on your backboard, it's a good idea to create enlarged versions, making them more visible and attractive. One year, we found an excellent solution to drawing endless lines with thick markers on construction paper, only to have them come out fuzzy. The art supply store had, in a final markdown basket near the door, rolls of glossy, colored tape in a variety of shades and widths. These can be used in either line or bar graphs for a professional and artistic look. Thin, black tape can be used to show gradations on graphs. Remember to make them accurate as well as attractive.

Photographs or illustrations can be mounted using hinges, corners, or glue. Be sure, that anything pasted or glued on your backboard will stay in place for the duration of the exhibit. If you, a friend, or relative are artistic, line drawings can also be an eye-catching addition.

A good cement useful for bonding paper is the best way to paste things on a backboard. Buy the proper adhesive for the material you are using, and follow the manufacturer's directions, for example, to ventilate the room properly when working with certain types of glue. Also investigate using staples, nails, push pins, or other methods of attaching material to your display.

At this point, it's also a good idea to select the colors for the display. Contrasting colors that attract attention without being loud or garish usually work well, and metallics are good attention getters.

Fig. 8-13. Caleb Johns; *What Effect Does Drafting Have on Energy Requirements?*

Fig. 8-14. Kelly O'Neill; *Which Wood Burns Best In a Fireplace?*

Fig. 8-15. Soya Akira; *A Study of the Wasp—The Mechanism of Determination of the Sexual Function in Vespula Lawisii.*

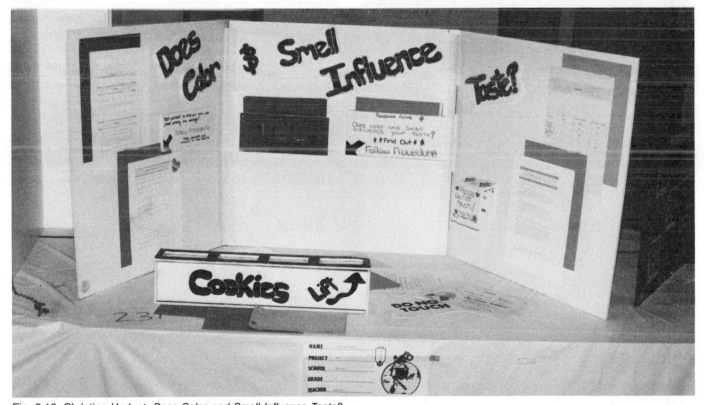

Fig. 8-16. Christina Hodnet; *Does Color and Smell Influence Taste?*

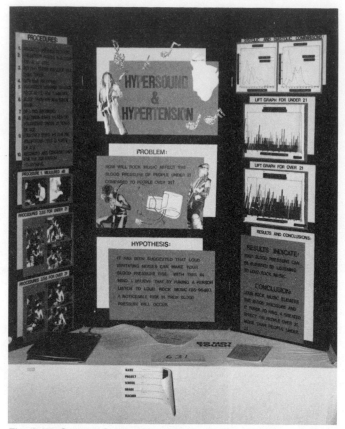

Fig. 8-17. Gregory Syrek; *Hypersound and Hypertension.*

Many students have also used colors that reflect the project itself, such as green for botany projects. In any event, if you're using several colors, check how they look together before making a final decision.

Always (no exceptions) have extra material on hand. A Sunday night, when the display is due first period on Monday morning, is the worst possible time to find that you've misspelled something on the very last sheet of construction paper. It's a mistake that has often reduced panicked families to rummaging through wastepaper baskets for an even halfway respectable scrap.

Careful planning and measuring, however, will reduce the need for such last-minute resourcefulness. Lettering seems to be the area most vulnerable to errors, so here are a few hints that are useful, regardless of whether you're using press-ons, stencils, or doing your own.

First of all, write out what you're going to say. Check the spelling of each and every word to be included. Decide on capitalization and punctuation, and check that for accuracy and completeness, too.

Next, check the number of lines the text will occupy. To do this, decide upon the size of lettering you'll use. Then, double check and make sure your message takes the same amount of room that you anticipated. Also verify that what you need to say will fit across the panel. If not, reach for a thesaurus to find alternate ways of stating the needed information.

If you're buying press-on letters, count how many of each letter, in each size, that you'll need for the entire display. Although this may seem like a tedious task, it may save money for buying extra sheets "just in case," or even worse, running out of E's at the last minute. If the worst happens, however, and you run out after the store has closed, remember that with a good eye, a steady hand, a sharp razor blade and fragments from unused J's and Z's, an F can become an E, a V can be transformed into an A, and an P can be made from an R!

When you're finally ready to do the lettering, to assure a professional looking job, take these steps:

1. Lightly draw lines across the sheet to be sure that your lettering is level.

2. Lightly mark the center of each line. Then, place the center letter(s) of your text on that spot and work outwards, until you've completed your line. If you're using stencils, you may wish to outline these in pencil and check the line before filling the forms with marker or paint. If, on the other hand, you're using press-ons, keep them *lightly* attached in case you need to remove them later.

Now that you're ready for the final steps, attaching written and visual materials, follow similar steps. Lightly mark exactly where the particular item is to go, and make sure it creates the effect you really want before gluing it into place.

Attach your notebook to the lower-left corner of the center panel and the display is done. Open up the backboard, step back and enjoy. You've done a great job!

Chapter 9

The Fair

The big day is approaching. You've done your best work, created an attractive presentation, and you're ready for your first science fair.

SETTING UP

On the day before judging, you'll need to set up your display in the exhibit hall. This area can be as simple as a school lunchroom or as elaborate as a civic center or convention hall. Regardless of where it's held, however, some things remain the same.

When you enter the exhibit area, your display will be checked for size, and then assigned a number, which will indicate your spot. Projects in the same category are always grouped together. This is for convenience, because there are often different judges for each category. Within a category, places on the exhibit tables are usually assigned by alphabetical order, so your location is the luck of the draw. The only exceptions are oversized displays or those needing electrical power.

Incidentally, project areas are usually guarded by professional security personnel. Don't worry, then, about your notebook or other display material getting "ripped off." Other exhibitors and their parents are also concerned about expensive computer equipment included as part of their display. Be assured that everything possible is done to protect your property.

Installing your project can be less hassled if you're prepared. If setting up involves crawling and climbing, wear old clothes. You may want to bring a small collection of tools, such as a hammer and nails, or a screwdriver, to fix anything that breaks in transport. A glue stick to fix corners of your display material that have become dog-eared might also be helpful. If you use light bulbs that can burn out or glassware

that can break, it's a good idea to have extras on hand for the duration of the science fair. Incidentally, having your own tool kit may make you the most popular person in the hall on set-up day.

JUDGING DAY

Understandably, you're nervous. You know your project is good, and your display is attractive but don't know what the judges are looking for. When you set up, you may have seen some other projects in your category. With a total lack of confidence, you may feel that everyone else's project is better than yours. Don't panic, though—opening night jitters are normal.

On judging day, most science fairs are closed to the public. Until the awards ceremony that afternoon or evening, it's just you, the officials and the judges. First, realize that the judges will be looking at many other projects on the same day that they're examining yours. One way of attracting positive attention, even *before* they review your work, is to appear organized and professional.

Here, I'm going to take a moment and sound a bit like your grandmother, but remember that her advice must have been good if it hung around for so long!

If possible, sleep well the night before. Fatigue and stress do not mix well when trying to make a good impression. Next, look well. You need not wear expensive clothes on judging day, but neat, conservative clothes (maybe the same ones you wore to Thanksgiving dinner) and a clean, well scrubbed look help create a positive image. Also act well. Don't eat, chew gum, clutch a soft drink, or slouch when the judges are walking through the exhibit area. Finally, speak well, both verbally and nonverbally. Smile and speak politely when questioned or addressed.

Fig. 9-1. San Diego Science Fair floor plan.

Although your project is being evaluated, not your manners or appearance, remember that the judges are only human. They'd rather spend time with an agreeable exhibitor than an unpleasant one.

It's also important to be ready to discuss your work. At some science fairs, each student is expected to give a short oral presentation about his or her project. At others, the judges will circulate through the exhibit area, looking at every project, and asking questions. When asked, don't wave your arm and say "it's all here on the backboard."

The judge already knows that. He or she wants to hear you express yourself.

Even if you're not expected to give a prepared oral presentation, it's a good idea to plan and rehearse what you will say. Check and practice the pronunciation of difficult words. Try to remember these helpful hints:

1. Throughout your discussion, remain friendly and polite. Maintain good eye contact, not staring at the floor or the ceiling. Stand up straight. (Hi, grandma!)

2. Be ready to introduce yourself and your project.

3. State your purpose and hypothesis clearly and distinctly, followed by a *brief* description of your procedures. If you have your specimens or apparatus on display, now is the time to point it out. Invite the judge to examine it more closely or try operating the device.

4. List your results. Now, point out any graphs, charts or tables, either in your notebook or on your backboard.

5. Discuss your conclusions, including whether or not the experiment proved your hypothesis.

At this point, even if you've given a prepared presentation, the judges will, no doubt, have some questions. Some common areas are:

—How or why did you get interested in the topic?
—Are there any aspects of the experiment or research

Fig. 9-2. Illustration of award ribbon.

that you might have changed or corrected, if you had the time?

—Do you intend to continue work in this area? If so, how? If not, why not?

—What practical applications or future use does your work have in "the real world?"

—Have you seen the article last month in the Bladedy-Blah Magazine by Dr. Such-and-such dealing with the further implications of etcetera and so forth?

First of all, don't be afraid to admit that you don't know an answer or that you haven't read the article or book in question. You'll make a far better impression with your honesty than with a futile attempt to "snow" an expert.

This brings us to another point. With either your answers or prepared speech, certainly try to be accurate and informative. However, don't try to give the judge all the information in your notebook in less than two minutes. Summarize, show and tell, covering the main points.

Science fair judges agree that the factors that come across most positively are knowledge and enthusiasm. A student who takes the path of least resistance by selecting a topic requiring a minimum amount of work is unlikely to rate high marks, even if he or she has a beautiful display. However, someone who has really worked to learn as much as possible and made an effort to follow good scientific procedures will impress the judges, even if the experiment has not worked out well. As we've all heard before, enthusiasm is contagious. Participants who are excited about their experiment and research show that they've gained the true benefit of doing a project and being in a science fair.

When the talk or the question and answer period is over, smile, shake hands and thank the judge. When he or she has moved on to the next backboard, you can breathe a sigh of relief. Now it's only a few more hours until you find out if you've won an award and the chance to advance to the next level.

AND THE WINNER IS . . .

That's right. You're the winner, before you even get to the awards ceremony. In fact, you've been a winner from the moment you started, as you selected your topic, completed your experiment, formulated your results and conclusions and created your display. Your perseverance in sticking with it through the difficult times, your flexibility in finding "alternate plan B" when your experiment seemed to be failing, and your creativity, curiosity, and talents have won you new knowledge and confidence, and perhaps even a life-long enthusiasm.

However, awards are nice! From the Oscars and Emmys to your local science fair, everyone wants the acclaim and recognition of their teachers, family and friends. There is usually a wide variety of awards. There are the awards of the particular science fair itself, which consist of first, sec-

ond, and third place in each category. Then, there are sweepstakes winners, who are judged to have the best overall projects. Finally, various private companies, research institutes, and military representatives give awards to projects that enhance the knowledge in their particular area. Such special awards are often conferred in such diverse areas as ophthalmology, nutrition, aerospace, and oceanography.

At the International Science and Engineering Fair, many such special awards are given. NASA, the International Aerospace Hall of Fame, the U.S. Armed Forces, corporations, such as General Motors and Eastman Kodak, and many others confer prizes. Trips to other science fairs, both in our country and abroad, are coveted prizes. There are also many scholarships awarded, among them, complete four-year scholarships to Ohio State. In the International Fair, the highest award, given to two projects selected from the Grand Prize winners in each category, is the Glenn T. Seaborg Nobel Prize Visit. This consists of an all-expense-paid trip to attend the Nobel Prize ceremonies in Stockholm, Sweden.

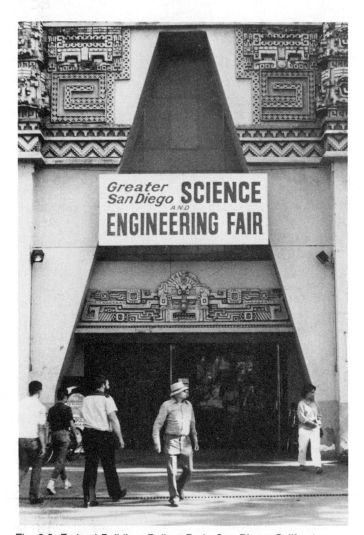

Fig. 9-3. Federal Building, Balboa Park, San Diego, California.

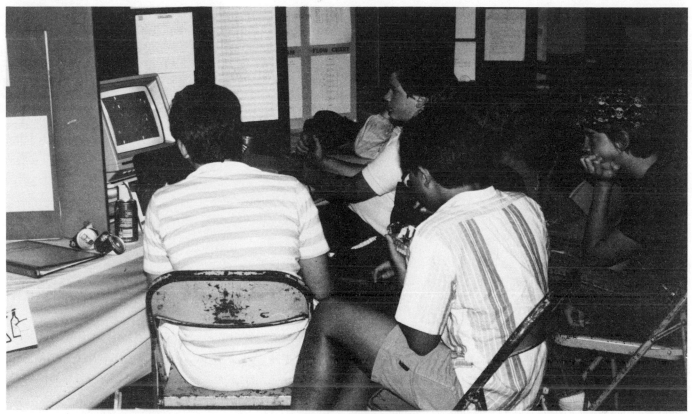

Fig. 9-4. Fun at the Fair!

If you've won an award, congratulations!! Hopefully, it's the first of many. But, if you didn't get a prize, don't let your disappointment spoil the science fair for you, or dampen your enthusiasm for future competition. If you're lucky enough to receive copies of your evaluations, you might get some good ideas on how to improve your project, and could possibly consider restructuring your project and entering it the following year. But, whatever you've come away with, you've gained a working knowledge of the scientific method and an insight into an area of science.

MEET ME AT THE FAIR

Once the excitement and tension of judging and awards night is over, enjoy yourself. Most science fairs go on for several days, to allow both school children and the public at large the opportunity to review the projects.

If you like meeting and talking with people, this might be fun, especially if you have the kind of project that generates many questions. Some exhibitors even prepare hand-outs describing their work to give out to interested viewers. Most students I've spoken to, especially first time participants, have spoken with pride of meeting an acquaintance who didn't even know they were in the fair.

You'll also get the chance to look at others' projects. For comparison, interest, or perhaps ideas on future projects, most participants enjoy looking at others work.

Finally, in many cities, museums, hospitals, universities,

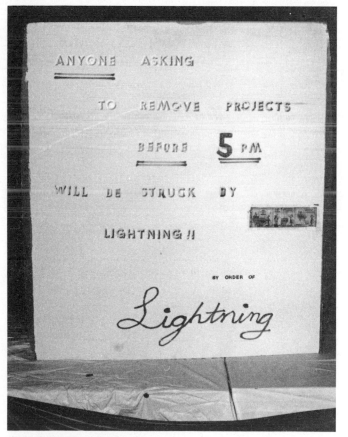

Fig. 9-5. Struck by lightning!

and other institutions join together to congratulate and celebrate the participants in their local science fair. In San Diego, California, the fair is held in Balboa Park, the site of most of the city's museums. For the duration of the fair, an exhibitor badge gains the student free admission to any museum. There are also lectures and guided tours, especially for participants. The University observatory, a behind-the-scenes look at the zoo, or a trip to a medical center, can be fun and informative. Counseling sessions, given by professionals to advise students about various scientific

careers, are also available.

You'll be required to keep your project on exhibit until the specified end of the fair, or suffer the consequences. This is to insure each student will check out his or her own project materials when the fair is over.

In the end, at 5 P.M. on Sunday, April 13, 1986, officials made the announcement "The 32nd Annual Greater San Diego Science and Engineering Fair is officially closed." The responding cheer was one of joy and victory, for the participants, their teachers and their families.

Appendix A

ISEF Project Categories

BEHAVIORAL AND SOCIAL SCIENCES

Psychology, sociology, anthropology, archaeology, ethology, ethnology, linguistics, animal behavior (learned or instinctive), learning, perception, urban problems, reading problems, public opinion surveys, and educational testing, etc.

BIOCHEMISTRY

Molecular biology, molecular genetics, enzymes, photosynthesis, blood chemistry, protein chemistry, food chemistry, hormones, etc.

BOTANY

Agriculture, agronomy, horticulture, forestry, plant biorhythms, palynology, plant anatomy, plant taxonomy, plant physiology, plant pathology, plant genetics, hydroponics, algology, mycology, etc.

CHEMISTRY

Physical chemistry, organic chemistry (other than biochemistry), inorganic chemistry, materials, plastics, fuels, pesticides, metallurgy, soil chemistry, etc.

COMPUTER SCIENCE

New developments in hardware or software, information systems, computer systems organization, computer methodologies and data (including structures, encryption, coding, and information theory).

EARTH AND SPACE SCIENCES

Geology, geophysics, physical oceanography, meteorology, atmospheric physics, seismology, petroleum, geography, speleology, mineralogy, topography, optical astronomy, radio astronomy, astrophysics, etc.

ENGINEERING

Civil, mechanical, aeronautical, chemical, electrical, photographic, sound, automotive, marine, heating and refrigerating, transportation, environmental engineering, etc. Power transmission and generation, electronics, communications, architecture, bioengineering, lasers, etc.

ENVIRONMENTAL SCIENCES

Pollution (air, water, land), pollution sources and their control, waste disposal, impact studies, environmental alteration (heat, light, irrigation, erosion, etc.), ecology.

MATHEMATICS

Calculus, geometry, abstract algebra, number theory, statistics, complex analysis, probability, topology, logic, operations research, and other topics in pure and applied mathematics.

MEDICINE AND HEALTH

Medicine, dentistry, pharmacology, veterinary medicine, pathology, ophthalmology, nutrition, sanitation, pediatrics, dermatology, allergies, speech and hearing, optometry, etc.

MICROBIOLOGY

Bacteriology, virology, protozoology, fungal and bacterial genetics, yeast, etc.

PHYSICS

Solid state, optics, acoustics, particle, nuclear, atomic, plasma, superconductivity, fluid and gas dynamics, thermodynamics, semiconductors, magnetism, quantum mechanics, biophysics, etc.

ZOOLOGY

Animal genetics, ornithology, ichthyology, herpetology, entomology, animal ecology, anatomy, paleontology, cellular physiology, animal biorhythms, animal husbandry, cytology, histology, animal physiology, neurophysiology, invertebrate biology, etc.

Category Interpretations

Below are project areas about which questions frequently arise. It is included only to provide some basis for interpretation of the category descriptions.

INSTRUMENTS

The design and construction of a telescope, bubble chamber, laser, or other instrument would be Engineering if the design and construction were the primary purpose of the project. If a telescope were constructed, data gathered using the telescope, and an analysis presented, the project would be placed in Earth and Space Sciences.

MARINE BIOLOGY

Behavioral and Social Sciences (schooling of fish), Botany (marine algae), Zoology (sea urchins), or Environmental Sciences (plant and animal life of sea, river, pond).

FOSSILS

Botany (prehistoric plants), Chemistry (chemical composition of fossil shells), Earth and Space Sciences (geological ages), and Zoology (prehistoric animals).

ROCKETS

Chemistry (rocket fuels), Earth and Space Sciences (use of a rocket as a vehicle for meteorological instruments), Engineering (design of a rocket), or Physics (computing rocket trajectories). A project on the effects of rocket acceleration on mice would go in Medicine and Health.

GENETICS

Biochemistry (studies of DNA), Botany (hybridization), Microbiology (genetics of bacteria), or Zoology (fruit flies).

VITAMINS

Biochemistry (how the body deals with vitamins), Chemistry (analysis), and Medicine and Health (effects of vitamin deficiencies).

CRYSTALLOGRAPHY

Chemistry (crystal composition), Mathematics (symmetry), and Physics (lattice structure).

SPEECH AND HEARING

Behavioral and Social Sciences (reading problems), Engineering (hearing aids), Medicine and Health (speech defects), Physics (sound), Zoology (structure of the ear).

RADIOACTIVITY

Biochemistry, Botany, Medicine and Health, and Zoology could all involve the use of tracers. Earth and Space Sciences or Physics could involve the measurement of radioactivity. Engineering could involve design and construction of detection instruments.

SPACE RELATED PROJECTS

Note that many projects involving "space" do not go into Earth and Space Sciences. Botany (effects of zero G on plants), Medicine and Health (effects of G on human beings), Engineering (development of closed environmental system for space capsule).

COMPUTERS

If a computer is used as an instrument, the project should be considered for assignment to the area of basic science on which the project focuses. As examples: if the computer is used to calculate rocket trajectories, then it would be assigned to Physics, or if the computer is used to calculate estimates of heat generated from a specified inorganic chemical reaction, then it would be entered in Chemistry, or if the computer is used as a teaching aid, then it would be entered in Behavioral and Social Sciences.

Appendix B

Summary of Projects

BEHAVIORAL AND SOCIAL SCIENCES

Interaction Vs. Back-up: Testing Two Theories of Mouse
Navigation by Ann Wood

This experiment was made to determine if mice, running through a T-maze to locate and reach food rewards, used a magnetic field as a backup to visual cues, or used both methods interactively. Although the test results strongly indicated the use of both methods interactively, the statistical level of significance, 0.05, was not reached in order to conclusively prove the hypothesis.

The Influence of Favorite Color, Sex, Age, Eye and Hair Color Upon Wardrobe Color, by Lan Do

In this project 419 Subjects, ages 5-70, completed surveys concerning their age, sex, ethnic origin, eye and hair color, favorite color, most popular color in wardrobe and eye and hair color of parents. Results indicated that color preference is sex related, with females preferring red, purple, and pink. The survey showed a preference for cool colors over warm colors as age increased. However, the survey results indicated that a person's favorite color had no relation to ethnic origin, eye color, or hair color. The most abundant wardrobe color is related more to the sex related preferences than hair or eye color, and is partially related to the stated favorite color.

Teenage Stress, by Kimberly Lea Quinlan

A study was done, using a questionnaire, in nine schools (four private and five public). There were a total of 378 responses. The analysis centered upon reasons for stress, which teens felt the most stress and how teens acted to relieve stress. Parents and tests were the major stressors. Public school students were less stressed than private school students, and females were more depressed or stressed by world problems than males. Alcohol and drugs were seen as responses to stress rather than their cause, but overall, private school students tended to eat while under stress and public school students used television and radio for relief. The results of the study correlated well with the ABC News survey of teenagers.

A Study of the Effects of Selected Carbohydrates on Alcohol Absorption and Behavioral Modification in Humans, by Catherine Jacobus

This experiment tested certain carbohydrates (fructose, glucose, sucrose, and saccharine) to determine if they had an effect on blood alcohol levels. Seventy subjects performed computer tests, sequence tests, dartboard tests and straight line tests. Blood samples were also drawn and tested for blood alcohol level, as well as glucose evaluation and SMAC profile. The experiment showed that glucose and fructose lowered blood alcohol and resulted in a better performance on the tests than saccharine. The sucrose group showed a higher blood alcohol level than the saccharine group. These results show a possibility that some carbohydrates serve as blocking agents that may modify alcohol absorption.

Mouse Appetite and Sound, by Michael Bruce

The object of this experiment was to determine whether sound affected the appetite of mice. The experiment was inspired by some classroom discussion dealing with subliminal advertising.

The subjects were divided into three experimental groups, one which was subjected to no unusual sound, one which was subjected to low frequency sound and one which was subjected to high frequency sound. The experimental groups were subjected to the sound for one hour each day during the experimental period.

Each day, an equal amount of food was given to each group. The next day, the amount remaining was measured before an entire measure was given.

At the end of the experimental period, the results showed that the control group ate less than either experimental group and that the group receiving low frequency sound ate less than that receiving the high frequency sound.

BIOCHEMISTRY

The Rabbit Test—Phases V & VI, by Elizabeth M. Robinson

This project was a continuation of earlier projects, which sought to control animal reproductive drives in the wild. Phase V of the project tested the ability of oral steroids to limit the reproductive capacity of adult male rabbits. In the experimental group, three of four males given steroids became sterile, while the fourth sired a healthy litter. Another experimental group was given human birth control pills. The results in this group showed that the human birth control pills produced a marked increase in male offspring.

A Comparison of Sugar Content in Soft Drinks by Jennifer L. Decker

This project tested our regular soft drinks and their diet equivalent, to determine the percentage of sugar in each. Coca-Cola, Diet Coke, 7-Up, Dr. Pepper, Diet Dr. Pepper, Pepsi, and Diet Pepsi were tested.

Hydrochloric acid was added to each sample to break the sugar down to reducing sugars. Benedict's solution was then added, and then standard glucose solutions were made in 5%, 10%, 15%, 20%, 25%, and 30%. Benedict's solution was added to these, as well.

Each solution was then heated, in order for the Benedict's solution to react. Each sample was then examined under the microscope, in order for Ms. Becker to count the black dots that represented the sugar. The number of dots in a 1/8 section were counted, both in the eight samples, as well as the standard percentages.

The results showed that regular soft drinks contain these approximate amounts of sugar: Dr. Pepper; 30%, Pepsi; 30%, Coke; 25% and 7-Up; 25%. No black dots were found in the diet drink samples.

The Synthesis and Biochemical Evaluation of Natural Estradiol Analogs in Cancer Chemotherapy by Latisha M. Keith

The objective of this process was to produce a synthetic estrogen analog. The application of this technology is important, because direct evidence has linked estrogen to the development of breast cancers. This study will continue with testing on live mice.

BOTANY

Radiation's Effect on the Plant Tissue Hormone, IAA by Jeffrey S. Morgheim

This experiment tested whether increased radiation (caused by a decrease in the ozone layer) altered the elongation properties of the plant regulating hormone IAA.

Oat seeds were germinated. An experimental group was placed in a solution containing IAA, which had been irradiated with laser radiation for 20 seconds. The control group was placed in IAA in its normal condition. Both groups were placed in the dark for 24 hours and measured for elongation. As the experiment progressed, the degree of IAA radiation in the experimental group was increased. The results did not indicate an increased elongation based on the increased radiation of IAA.

The Effect of Radiation on Bean Seeds and Seedlings, by John Stoffel

This project also tested the effects of radiation on seedlings. The experiment was carried out over several years. In the first phase, germinating and dormant seeds were exposed to varying degrees of radiation. The next phase exposed the buds to 6000 rads of radiation and in the final phase, the seedlings were exposed to 2000-6000 rads, at 7, 12, and 21 days of growth. Results showed that higher doses inhibit growth at a 2:1 ratio, but that wet seeds were twice as affected as dry seeds. The greatest effect appeared to occur at the period of greatest growth, 8-16 days after planting. The portion of the plant most affected was the terminal bud, followed by the roots and then the stem.

Mr. Stoffel concludes that this may be due to the rates of cell division at that time. He further concludes that due to the effects of radiation effects, in case of nuclear accident, knowledge of effects on plant life is essential to understanding the possible survival of the planet.

The Effect of Seed Size on Wheat—A Quantitative Study by Neil Ireland

This project attempted to measure the effect of seed size on the success of the crop. This is important, because wheat is a major food source throughout the world.

1000 seeds were measured with a micrometer and then categorized in 10 evenly spaced groups. Ten from each group were then grown in cotton wool in petri dishes and watered with a mineral solution. This was repeated for each of the experimental groups.

The number of seeds germinated, as well as the average

height of the groups of plants, were recorded. In addition, grazing was simulated by cutting off the plants at a specified height. Frost was simulated by freezing the seeds, and drought was simulated by water deprivation.

The results showed that a larger seed increased the initial growth rate and regrowth rate after grazing, as well as drought and frost resistance.

CHEMISTRY

Comparative Study of Raw and Pasteurized Milk from Different Dairies of Puerto Rico, by Francis Vazquez

This project was undertaken to determine the degree of milk contamination from different dairies, and the cause of such contamination. He tested both raw and pasteurized milk in quality control tests to determine acidity, lactometry and total solid percentage, as well as evaluating taste, odor, and appearance. The tests were also used to detect and measure the presence of bacteria and antibiotics.

The results showed that the consumers received milk with a 71% of quality. He determined that environmental qualities were responsible for 29% of the impurities. The presence of antibiotics, mainly penicillin, was probably due to penicillin used in the treatment of cows. Research was carried out using the facilities of the Puerto Rico Dairy Products Association.

Stop—You May Be Using The Wrong Detergent, by Gwen S. Pearce

This experiment tested eight detergents to determine the most effective in removing stains of grass, combined food, mud, and chocolate stains. All stains were cleaned in a water solution of each detergent and then evaluated and graphically represented. One detergent was rated best in terms of stain removal, yet conclusions were based on qualitative rather than quantitative analysis.

Nutra-Sweet in Diet Soda—A Study of Storage Temperature by Randall Shank

This project sought to study the degree of breakdown of Nutrasweet (Aspertame) in solution form. The experiment used diet soda samples, stored for a trial period of ten weeks, at 5, 15, 25, and 25 degrees C. Measurement was made using a High Performance Liquid Chromatograph.

The aspertame quantities decreased significantly at higher temperatures, with changes reflecting a logarithmic relationship. The data led to a formula that was used to produce data that closely followed the experimental data. The main "discovery" of the project was actually the formula, which may prove useful for determining the long term safety of aspertame.

COMPUTER SCIENCE

Computerized Speech Analysis & Recognition, by Carl T. Donath

This project was inspired by the need to input data into a computer without the need to type. The objective was to design an electronic circuit that could convert sounds into a series of numbers. First, the sounds, which are really changes in air pressure, are converted to electrical voltages via a microphone, and then turned into numbers by an analog-to-digital converter. Once the circuit was working, a program was written to interpret the numbers, searching for something that would distinguish one word from another. He finally drew an "envelope" around the graph that the word represents. These "envelopes" are stored in memory, and then used as the basis of comparison to subsequent "words" that are input.

Textlink by David Calabrese

This project developed the TEXTLINK system, which can "read" material from a typed or printed page and enter it into the computer memory. The purpose of this was to eliminate the need to re-key material for new word processors, data bases, etc., at a reasonable cost.

TEXTLINK works by using structural character recognition to identify characters within a computerized snapshot of a page. Digitizing equipment, which is relatively inexpensive, is used to take the photograph, "gives" the image to software, which recognizes the letter based on its unique structural components. Therefore, TEXTLINK achieved its objective by being an accessible, cost-effective text reader.

Computer System for The Handicapped, by D. Scott Smith

This project also attempted to use a computer system with "off the shelf" software geared for the handicapped. The object was to input data either with a speech synthesizer, or a joystick and/or touchpad. Three software programs were used, *English Based Interactive Language* (EBIL), a super high level language; *Words*, simple word processor; and *Talker*, a speech synthesizer.

Scott tested these programs over eight months, without using a keyboard or monitor, and found that the system could, in fact, be effectively used with alternate input and output devices.

Mendelevia by Angel de la Cruz

The purpose of this project was to create a computer program that would teach some of the basics about the elements. The basic assumption was that most of this material, such as names, symbols, atomic weights, etc., was rather boring.

Therefore, the objective was, rather than creating educational software, to write a "game" program, which students would enjoy, and therefore continue playing and learning chemistry in the process.

EARTH & SPACE SCIENCES

The Shaking Earth, The Burning Sky by Mike Iritz

This project attempted to correlate the incidence of earthquakes of magnitude 6.0 and above, and 7.5 and above, with incidence of high solar activity. Data for the years 1937 through 1975 was collected, graphed, and correlated using the Coefficient Correlation. Although graphically, a correlation was shown, statistically, the hypothesis was disproved.

The Effects of Microgravity on Embryogenesis by John Chin-Hung Leu

This project studied the difference in the development and spawning of amphibian eggs under simulated hypogravity conditions. The experiment showed that in the control group, more than 60% of the eggs were shown to develop normally with respect to size, form, etc. However, in the experimental group, development of the eggs was delayed, and also structure of the eggs was abnormal. Mr. Leu plans further studies to analyze the precise factors that lead to the developmental delay, and examine the synthesis of chemicals within the eggs.

Variation in Crystal Habit of Pyrite from Otoge Mine, by Koji Morimoto.

This project studied the different patterns of growth of crystal formations on pyrite taken from different portions of the same mine. About 20,000 samples of pyritic crystal were collected and classified. After analysis and study of the samples relative to their source, the conclusion reached was that the location of its formation, particularly as rates of cooling varied in different parts of the mine, was a fundamental cause of the variances in rate and type of crystallization.

Hohokam Pottery Porosity, by Andrew J. Wolf

This project involved testing the porosity of pottery shards, to determine if this study would indicate that pottery was produced differently depending on its intended use.

The hypothesis was that the pots would be more porous because they were used as water containers, and a more porous container would permit a film of water to form on the outside, creating an evaporating effect. The test was done by weighing the shards dry, then submerging them in boiling water and weighing them again.

The testing proved the hypothesis incorrect. Mr. Wolf speculates that this may be because bowls may have been used for cooking, as well as for storage, and that in cooking, more porous materials would heat more evenly.

ENGINEERING

Design, Construction & Operation of a Continuous/Pulsed Ion Beam Accelerator, by Stephen M. Jacobs

The objective of this project was to build a small accelerator for use in high energy particle physics. The idea was derived from previous projects, which dealt with various aspects of the atom, namely the electron cloud and the nucleus.

Materials for the project, such as small parts and electrical supplies were purchased at shops, although a great many more specialized pieces of equipment were borrowed. Household materials were also used. The purpose was to develop the best design, using "common" materials, for safe and efficient use.

The main problems encountered were the proper focusing of the ion beam and the minimizing of the gassing of the device.

Fly Ash Foam—A Method to Create A Ceramic Material from The Waste Products of Coal Incineration Through The Use of Microwaves by Caroline K. Horton

The objective of this project was to create a useful product, a strong lightweight foam, from the fly ash, which was a waste product from coal-fired heating and electrical generating facilities.

To produce the foam, Horton developed a solution to add to the fly ash and then microwaved it, producing a strong ceramic foam. To shape the foam, she produced fiberglass molds in which to encase the foam solution prior to the microwave process. Horton began her research at the Los Alamos National Laboratories, which produced its own fly ash foam. However, the foam produced by her method was stronger and more durable than that developed at Los Alamos. Therefore, because of its commercial application in construction and insulation, she is currently applying for a patent for this new method of producing the foam.

Coil Beam Magnetic Accelerator, by Michael James

The objective of this project was the development of a coil gun, which fires at great speed with low input power, by use of induced magnetic fields. This project was triggered both by previous research as well as an interest in SDI research.

It was determined that to yield high input velocity and low input power, large capacitors would need to be used to supply the magnetic coils with current. Also, the wire gauge, length of coil and number of wraps were optimized to supply highest acceleration. Optical triggers were used to fire each stage, making the systems self-correcting. Five coil stages yielded a projective velocity of 28.68 meters per second.

ENVIRONMENTAL SCIENCES

Rehabilitation of Ground Water Aquifers, by Terri Newport

This project grew from a concern with hazardous waste sites, particularly how they affect ground water. The objective of the project was to develop and demonstrate methods of identifying, isolating, and rehabilitating ground water aquifers.

Newport accomplished this using five physical models that would duplicate actual conditions. These models were constructed of plexiglas, to permit easy observation, and fitted with injection and pumping wells, which allowed for sampling the contamination concentration. In each instance, a rehabilitation strategy was developed and demonstrated, identifying its effectiveness and limitations.

This project grew out of work on a prior year's project, Ground Water Pollution—Identification, Isolation and Rehabilitation.

Acid Rain—Causes, Effects, and Cures, A Seven Year Study, by Leslie S. Thomas

This seven year project attempted to find the causes and effects of acid rain, as well as a method of correcting the damage inflicted by acid rain. The final phase dealt with a method of implementing corrective measures previously discovered. The neutralizing agent was tested in two acidified ponds in the Adirondaks. Once the ponds were neutralized, they were restocked with native fish and plant life. After eight generations, these species are still thriving.

Controversial Issue—Are EPA Toxic Waste Standards Safe? by Eric J. Felt

This project attempted to prove whether the EPA's safe levels of toxic waste were, indeed, safe. To conduct the experiment, Mr. Felt used the acceptable levels of three different types of toxic waste, herbicides, solvents, and heavy metals. He then tested planaria for their ability to regenerate as well as perform conditioning tests.

Although each of the wastes produced different impairment levels, using the Chi-square method of validation, it was shown that the "acceptable" levels of each waste, did, in fact, affect the regenerative and conditioning abilities of planaria.

Can A Charcoal Filter Reduce The Effects of Automobile Exhausts on Chick Embryos? by Libby Haines

This project was based on prior research, which proved the harmful effects of fumes, from both leaded and unleaded gasolines, upon chick embryos. In this project, a filter was developed in a attempt to lessen the effects of the fumes, using plastic pile with charcoal. These filters were placed on cars with leaded and unleaded gasoline, and then introduced into zip-lock bags containing chick embryos. Unfiltered fumes, as well as a control group subjected to no fumes, were also part of the test.

The results showed increased growth of the embryos exposed to the filtered fumes, as opposed to those exposed to non-filtered fumes.

An easy, inexpensive method of reducing pollution is the possible development of a "throw-away" charcoal filter to be attached to the end of a gasoline exhaust pipe.

MATHEMATICS

Blackjack by William Chen

Chen's project attempted to expand on or improve the various blackjack systems available, which were based on counting cards. He determined that the existing systems yielded a return ranging from 1.1% to 1.3%. He altered Julian Braun's system to count aces, but was unable to appreciably improve the rate of return. Mr. Chen concluded that the only effective systems were based on counting cards, and even when those systems are adhered to, gain is at a slow rate. The tests were run using Turbo Pascal.

An Original Method for Solving Tangential Circle Systems by Jennifer Wood

This project attempted to find the solution for a fourth tangent circle in a system when the other three were given. Although there have been solutions using quadratic equations, they were very lengthy processes. Ms. Wood's solution is shorter and more efficient. She developed a numerator, which is the product of the first two radii divided by the third, then normalizing the results. Subsequently, all additional circles could be found by sequentially adding all denominators to the sequence.

Wood's results were reviewed and verified by a professor of mathematics and engineering at the University of Southern California.

It's Not Cartesian—A Study of A New Graphing System, by Christopher C. Finger

Finger developed the Triangle graphing system in which each pair of coordinates defines a segment that joins the axes. This is unlike the Cartesian system where each pair of coordinates defines a point. The research for this project included studying the slopes of segments, conic sections, and inverse equations, as well as solving simultaneous equations to determine when segments cross.

Proceeding with the study, Mr. Finger treated the slope

of a segment as though it were a Cartesian line and proceeded to study the slopes in various equations. Study and research pointed out two flaws in the system. First, any segment with one coordinate of zero cannot be seen because it lies on an axis. Secondly, a step between x values must be specified, because with an infinite step, single segments cannot be seen.

MEDICINE AND HEALTH

Malaria Vectors—Eight New Species Implicated by Elke Baker

This project grew out of Baker's interest in tropical diseases and the opportunity to work in a tropical medicine laboratory over the summer. She was trained in the laboratory techniques while there and worked closely with a supervisor on many projects.

The objective of the project was the identification of the species of mosquitos that are infected with the human malaria species and the transmission in a particular geographic area. By testing the ELISA (enzyme linked immunosorbent assays), which detects sporozoites in mosquitos, on samples of 13 species of mosquitos, Baker uncovered the possibility that 8 new species of mosquito were implicated in the disease's transmission.

One advantage was Baker's research at the Walter Reed Army Institute of Research. A disadvantage, however, was that her specimens were two year old freeze-dried mosquitos, without accurate records of how, when, and where they were collected.

Immune Evaluation of HHT by Lisa J. Chin

HHT is a cephalotazine ester from several widely spread evergreen trees in China. Since HHT is under investigation for treatment in some types of leukemias, this project attempted to determine whether HHT was an immunosuppresive antitumor agent. Concentrations of HHT were added to a culture that contained human lymphocytes. All concentrations of HHT tested were found to be inhibitory to lymphocyte proliferation.

Stuck on You—A Study of Dental Adhesives by Stuart Allen

This project attempted to prove that substances that are high in sugar, such as cola or Kool-Aid, weaken the bond between teeth and braces, or other orthodontic appliances. To test this hypothesis, four groups of false teeth with orthodontic bands cemented on were soaked in four substances, cola, Kool-Aid, beer, and water. They were then subjected to 2250 grams of pull using a spring scale. Only one of the bonds came apart, on the first day of the experiment. Therefore, the hypothesis could not be proven by the experiment.

Computer Simulation of The Plasma Membrane's Transport Functions, by Dennis Petrocelli

This experiment attempted to create a computer simulation of the transport function of the plasma membranes. Two BASIC programs were created to accomplish this. The Develop program designed membranes by adding functional units to a lipid bilayer and the Membrane program simulated diffusion and facilitated diffusion across the membranes created with Develop. To validate the process, four experiments were done that investigated different aspects of transport. After the experiments, the simulations proved to be accurate, and therefore of potential use in the analysis needed to cure human diseases of the membrane.

The Canine Allergy-Obesity Connection by Missi J. Wilkenfeld

This project attempted to determine whether allergies predisposed individuals to obesity. The hypothesis was tested using a sample of 50 dogs. A set of criteria was developed to indicate an allergic condition, and the dogs were classified as allergic or non-allergic based on the results of the questionnaire. Six of the dogs were classified as normal and 44 were considered allergic.

Each dog's body weight was then compared to established norms. A pinch test was also done on each dog to determine the body fat content. If the dog's weight was 15% above average, it was classified as overweight, and if it was 20% above average, it was classified as obese.

Testing showed that the majority of dogs classified as allergic had weight problems. Of the 44 allergic dogs, 30 (68%) were either overweight or obese. Of the six normal dogs, only 2, representing 33%, were overweight or obese. Wilkenfeld hypothesized that the allergies slowed the metabolism. Hence, the results of this experiment indicated that allergy treatment might alleviate weight problems.

MICROBIOLOGY

The Curing Effect of Ultraviolet Light on The F'Proline + Lactose + Strain of Escherichia Coli, by Jessica Wertlieb

Wertleib was introduced to this topic by a professor of genetics at San Diego State University. Her main sources of information were the professor himself and several local university libraries. This project was of particular interest because although it was successful at winning several awards, Wertlieb overhauled her procedures at least seven times in the course of the experiment, because there were problems with faulty equipment, as well as difficulties in getting the organisms to grow properly.

The Differentiation between Human and Cow E. Coli, by Robin Rowe

Escherichia Coli is used to indicate fecal pollution when

fresh water is tested. However, when this substance is found, there has been no practical method of distinguishing between human and cow E.coli, therefore making it difficult to determine the cause of the pollution, and hence take steps to curtail it. This research consisted of two phases, one which tested the E.coli in five different sugars, and a second phase in which the sugars, plus brom cresol purple broth was used. The procedure used in phase two enabled absolute differentiation between human and cow E.coli.

Effect of pH Upon The Culture—Penicillin Notatum, by Shawn Bray

Studies show that penicillin notatum colonizes best at a pH of 7.0, and that growth is difficult above pH 8.0 or below pH 5.3. Because the penicillin mold grows on bread, yet can only colonize within the stated ranges, this project concludes that mold can be prevented by adjusting the pH level of bread.

Jalapeno Peppers, Garlic and Sage as Natural Preservatives and Germicidal Agents by Tracy M. Holje

This project tested whether a home remedy, consisting of jalapeno peppers, garlic and sage, work as a preservative on meat, and also can be used as a germicidal agent on cultures of mold and bacteria.

Both chemical and home remedies were applied and then bacteria was added to nutrient agar on the plate. After 24 hours, the samples were examined and the bacterial colony count recorded. The tests were repeated several times, each time, isolating the best preservatives and germicidal agents for further testing.

Finally, benzoic acid was proven to be the best chemical preservative, and whole garlic the best natural preservative.

PHYSICS

Obtaining Kinetic Energy through Heat, by Billy Scott

The objective of this project was to prove the theory that the temperature of a body is equivalent to the average kinetic energy of its constituent particles of molecules. The hypothesis was that a change in the temperature of the body would result in a change in the kinetic energy. Scott experimented with stretched opposing elastic bands between two points with equal and opposite force. When one side of the band was heated the entire band contracted, stretching the corresponding band an equivalent amount, hence resulting in increased energy. As long as the bands were kept in motion, the potential energy created was changed to kinetic energy.

To conduct the experiment, a device consisting of a bicycle wheel with existing hardware removed was used and elastic bands were symmetrically placed throughout the circumference of the wheel. A light bulb was used as the heat source.

What Happens When The Length of A Pendulum Is Increased, by Nicola Kean

This project examined the gravitational differences resulting in the different lengths of a pendulum. This project was interesting because Kean looked for a simple project with easily interpretable results, yet needed to delve into research in an unfamiliar field. However, she felt that she learned a great deal about physics and the laws of gravity as a result of the project.

In addition, Kean used a computer for her display, including graphs, which generated attention at the various fairs in which she exhibited.

Visual Reversal of Photographic Negatives, by Maryanne Large

The basis of this project was the analysis of obscure optical effects when black and white photographic film was viewed from the same side as the light source. These effects were the scattering effect, resulting in a positive image and a reflectance effect, producing a negative image. Testing was done to quantitatively determine the effects of density development and the angle of incident light on the images. Qualitative analysis was carried out using colored filters and electron micrographs were also made to aid the understanding of the importance of the structure of the film itself to the reflectance negative.

The project was inspired when Large accidentally noticed the optical effects when developing and printing photographs at home.

Real-Time Transmission of Holographic Interferometry-Determination of Minute Refraction Index Changes for Fluid Aerodynamic Applications, by Mark David Owens

The objective of this research was to design, construct, modify, and apply a real-time transmission holographic interferometry system, capable of measuring minute changes in refraction indexes to facilitate the visualization of water flow past a test specimen, thus showing the water's relationship to the object.

Most of the device was built using scrap materials, for example, an old water pump from a washing machine. Over forty devices were built and tested, until one was found to satisfy the project objective. This design may act as an alternative to standard water channels and wind tunnels for military, medical, chemical, and engineering applications.

What Is The Best Concave Design for Water Skis, by Ronald E. Thompson

First, Thompson determined that the concave design

for water skis was superior to flat design, but also discovered that there hadn't been any design innovations. He then created over 30 new concave ski designs, then tested them to narrow the selection to two designs.

These two were then tested against the concave ski design manufactured presently, all 1/5 scale size. Tests included surface tension, friction, and stability, conducted at a speed of 22.5 mph.

One of the designs proved superior to those currently in use, enabling easier learning and proficiency with the sport. A secondary product of the research applied the design to boat hulls, creating one combining the strength of a deep V hull with the speed of a tunnel hull.

ZOOLOGY

The Role of Amoebocytes in Zinc Uptake and Transport in The Snail Helix Aspera, by Ben Cheng

The objective of this experiment was to determine the role of mobile cells termed amoebocytes, which are amoeba-like cells, in the uptake and transport of metal particles in molluscs. These cells can accumulate these metal particles and transport them to other organs, such as the kidney.

To determine the role of amoebocytes in this process, they were isolated from the blood of the land snail, Helix Aspera, and then incubated with 65-zinc, because it plays an essential role in biological systems, and 109-cadmium, which does not participate in any known biological processes. The partitioning of the metal between plasma and amoebocytes was monitored in order to determine the uptake of soluble metals by ameobocytes.

The results showed that the amoebocytes apparently do not play an important role in the uptake and transportation of zinc and cadmium in the snail, Helix aspera.

Carbon Dioxide Is Necessary for Coral Growth, by Daniel S. Cheng

This experiment was conducted with two aquariums, one containing fish and one without fish, each containing coral growth. Observations and measurement on the amount of growth of the coral was made over a period of 14 days. After analysis of the data, the conclusion was that carbon dioxide, in the experimental case, generated by the fish, was necessary to the growth and survival of the coral. This was a rather expensive project, requiring the purchase of two aquaria plus four pumps. Mr. Cheng also encountered some difficulty collecting identical species and sizes of coral for the experiment.

A Study of The Wasp—The Mechanism of Determination of The Sexual Function in Vespula Lawisii, by Soya Akira

This project attempted to determine why queen wasps can oviposit, or lay eggs, while workers, which are also female, cannot. There are some basic differences in the reproductive organs, as follows. First, the ovary of the queen develops well, while those of the workers do not. Second, the ollicular egg cells of the workers do not grow as powerfully as that of the queen. Finally, oocytes with yolk granules can be observed in the queen, but not in the workers. These differences were determined by the differences in the amount of hormonal secretion in the queen and the workers. This was the most important factor in determining the sexual function of the queen wasp and the workers.

Drugs Weave Patterns, by Doug S. Fay

This experiment showed the effects of caffeine, alcohol, and nicotine on the geometric pattern of a spider's web. Because these spiders build webs daily and consume them at night, dilute solutions of the chemicals were misted on normal webs, and photographs were taken the next day when new webs had been woven. Drug influenced webs differed drastically from normal webs, indicating that the drugs affected the spider's ability to weave its web, and hence, its ability to function normally.

Appendix C

Glenn T. Seaborg Nobel Prize Visit Awards Winners' Abstracts

37TH INTERNATIONAL SCIENCE AND ENGINEERING FAIR, FORT WORTH, TX., 1986

A New Algorithm for Optimal Route Selection, Pamela S. Heady

In the future, robots will become more mobile. However, to be really useful, they must be able to execute optimizing commands. The Travelling Salesman problem involves such an optimizing command. It requires finding the shortest simple path connecting a given set of destinations. An N-station Travelling Salesman problem has $(N-1)!$ possible paths, and for N greater than about 20, even the fastest computer cannot do an exhaustive search in a reasonable time. By using the most advanced search algorithms, problems with about 30 destinations can be solved, but still only when run for weeks on a Cray computer.

My goal was to find a very fast, very accurate approximate solution to the Travelling Salesman problem. I hypothesized that useful results could be obtained by eliminating the requirement of reaching an exact solution. This is a practical goal because in many situations a path that is only slightly longer than the best will be considered adequate.

By using geometric analysis I found four principles which allow 30 point problems to be solved in about 7 seconds. The 48 state Travelling Salesman problem was solved in about 14 seconds. For N less than 150, my program runs in polynomial time (exponent = 1). Where I could check the solution, it was always exact, but I can show that certain cases will not be precisely correct.

I call this a "robotics solution," because it is much more accurate and far faster than what a human can do, and can be programmed into a small, mobile computer.

36TH INTERNATIONAL SCIENCE AND ENGINEERING FAIR, SHREVEPORT—BOSSIER CITY LA., 1985

A Three-Year Study of Computerized Differential Identification Using the Dental Configurations and Bitemarks of Homo Sapiens, Sheryl L. Ames

Because fingerprints cannot always be used to identify individuals, modern science and technology is now looking at the teeth to provide a new means of differential identification that will be a more permanent record of an individual's identity. The purpose of this research is to develop an inexpensive and reproducible computerized system that utilizes the dental bitemarks of an individual for differential identification. A bite wafer was developed to register an individual's bitemarks and the bitemarks of 183 test subjects from elementary schools were taken. A computer program was then written to convert the bitemarks into numerical values that are used to identify unknown or missing persons. A comparison of the differences in the bitemarks of the test subjects after one year demonstrated the uniqueness and validity of these values as differentiating characteristics. Because of the uniqueness and relative accuracy of this system, several identification agencies and professionals in the field of forensic dentistry have expressed an interest in the system.

The Development of a System for Acquisition of 3-Dimensional Images, Eugene Sargent

Sight is our most relied on sense. We use it in almost

everything we do. In order for a device possessing artificial intelligence to be able to interact with its environment as we do, it would need some way of seeing the things around it. It would need a system for getting three-dimensional images of its surroundings.

In my project, I developed a system, to be used by a computer, for getting digitized images of three-dimensional objects. My system utilized a small neon laser and a system of motors and mirrors to scan the surface of the objects. After a great deal of experimentation with photodetectors, I developed a successful array of photodetectors to gauge the distance of the laser spot. It utilized three specially modified cadmium sulfide cells and a 216 fiberoptic light guide to collect and distribute the scattered laser light. I designed and built a circuit to process the outputs of the Cds array, which I interfaced to a Commodore 64 by adding three input ports to the expansion port of the computer. I designed a program to decode the output of the array and to store it on disks. I developed a system of assembly language graphics routines, which plotted the image in simulated 3-D and removed the hidden lines from the plotted image.

Although my system is still a prototype, it has produced surprisingly good results. It has a fairly low resolution and a limited range, but the data is consistent and free of noise, I plan to improve the recognition and accuracy of my system and to go on to image recognition.

35TH INTERNATIONAL SCIENCE AND ENGINEERING FAIR, COLUMBUS, OH 1984

Net Transfers of Lipids Between Plasma Lipoproteins: A New Biochemical Index for Detecting Human Coronary Artery Disease, Ann R. Davis

Purpose. To devise a new procedure for measuring simultaneous in vitro net transfer pathways of lipids between lipoproteins of human fresh plasma and to determine whether any differences in these pathways occur in middle-aged males with coronary artery disease compared to age-matched normal healthy male humans.

Procedure. Following informed consent, 10 ml blood samples were obtained from 12 middle-aged male patients with coronary artery disease and 12 age-matched normal male human volunteers. A new procedure was developed for measuring the extent of net transfer of unsterified cholesterol, esterified cholesterol, and triglycerides between human plasma lipoproteins during a four hour incubation of fresh plasma at 37 degrees C. This new procedure utilized cellulose acetate electrophoretic separation of plasma lipoproteins coupled with presently developed new "sandwich" methods for enzymatic staining of plasma lipoprotein-lipid bands.

Results. A new biochemical index number for coronary artery disease, named the plasma lipoprotein-lipid transfer (PLLT) index, has been calculated from six plasma lipoprotein-lipid net transfer pathways, which were found to be statistically different in middle-aged male patients with coronary artery disease compared to age-matched normal healthy male volunteers. The PLLT index averaged 9.27 + 2.38 (or 9.27 − 2.38 mean + or − S.E.M.) in 12 male humans with obstructive coronary artery disease and 0.54 + 0.25 (or 0.54 − 0.25) in 12 age-matched normal male humans, corresponding to a 1,617% increase in coronary artery disease ($p < 0.005$), with a 92% correct classification for the diagnosis of coronary artery disease.

Conclusions. The present data suggest that the PLLT index may have a possible use as an initial screening marker for coronary artery disease.

Immunity to Herpes Simplex Virus I Conferred by Genetically Restricted Variants and Immunoactivating Factors, Timothy A. Thrailkill

Active immunity to herpes simplex virus 1 has been conferred with the use of an experimentally effective live vaccine. Vaccine samples were composed of inductor-exposed variants of parental H166 + in aerosolized form for oral inhalation. Variant populations were assured of monostrain purity by monoclonal antibody selections. Variants were screened for in vitro avirrulence, in vivo apathogenicity, antigenicity and immunogenicity. A prototypical vaccine affirmed sufficient attenuation and a potentiated vaccine confirmed immunogenic potential. Studies of nodal infectivity indicate primary replication in the oral cavity, with associated neural invasion, and secondary replication in splenic and helatic tissues. Proposed genotypes based on phenotypic data gave insight as to mechanisms of the abortive viral infection.

Final vaccine experiments explored immunizing regimens, Immunogenicity indices were designed on nomographs. Protocols on murine subjects resulted in full protection versus latent recurrence. Protection versus latency was by 1:1 cellular colonization in the trigeminal ganglia. Immunity was polytypic in nature in respect to other HSV-1 strains. Potentiated Cai 166 samples were safe in immunocompromised mice. Rabbit testing demonstrated the effectiveness of the vaccine. Overall results support the hypothesis that this vaccine is ready for human subject trials.

Recent studies indicate nontransmissibility of variants from vaccinated to unvaccinated individuals, probably due to reduced chance of morbid recrudescence by variants. Communal efficacy of the PL166' vaccine was proven resistant to acquiring contagious infection. Future research involving occult latency is in progress.

34TH INTERNATIONAL SCIENCE AND ENGINEERING FAIR, ALBUQUERQUE, NM, 1983

A Continuing Analysis of Alligator and Crocodile Tooth Microstructures and Their Potential for Evaluating Higher Systematics, Laura Trauth

Although naturally shed alligator and crocodile teeth are

some of the most commonly found Cretaceous fossils of North America, they have little taxonomic importance unless they are still lodged within a much rarer fossil skull and identified through skull fenestration. Results of a previous phase of this project suggested that internal ring structures of such teeth might be genetic invariants and, thus, have taxonomic value.

This phase of the project devises a means by which individual teeth may be quantitatively identified through their dentine ring systems, making the presence of a skull unnecessary. The teeth used in this study were known modern teeth from two families of crocodilians and unidentified fossil teeth. The teeth were evaluated mathematically, with the equations describing the curves of the dentine rings as the discriminating factors. The equations were found to differ at the species level for modern teeth. The fossil teeth were less easily classified because of their unknown character and interfering subring structures.

These structures were found to have possible counterparts in modern teeth soaked in a protease solution. The fossil teeth and the protease treated modern teeth were less fracture resistant than non-treated modern teeth—suggesting that protein is a strength factor. This hypothesis requires further research.

Thus, as well as providing a model for identification of individual crocodilian teeth, this study suggests that some fundamental aspects of tooth structure may be better understood through study of tooth proteins.

Tip Vortex Propulsion/A New Approach, Jonathan Santos

The tip vortex is a tornado-like disturbance, which is formed by the lower pressure lift zone drawing in the higher pressure air around the wing tip. I was curious to research possible ways to utilize the vortex instead of wastefully trying to eliminate it.

Researching this, I read an article on the unusual outstretched wing-tip feathers on larger soaring birds, and how those feathers detect the vortex for propulsion. I started my project with the idea of simply designing similar fins to be applied to the aircraft for the reduction in fuel consumption.

After three months of research in a home-made wind tunnel, I determined that this type of system would be inefficient and impractical due to the fact that the sails (except the first) lie in the turbulent wake of the sail before it. I then devised a way to incorporate the several separate fins into a single lifting surface. The technique to place this single sail at its most efficient angle of attack around the three-dimensional vortex was experimentally derived.

My concluding experiments determined that the "monosail" produces increased flight efficiency of 27 percent.

Glossary

abstract—A short summary of the main points of a project. This is normally between 200-250 words in length.

analyzed data—Data derived from raw and smooth data, from which conclusions can be drawn.

conclusion—Interpretation based on outcome of results and answering the question or comparison suggested by purpose.

control group—Identical to the experimental group in all aspects, except that no variables are applied. This represents the test group that has all variables standardized and forms the basis for comparison.

controls—This represents factors that are not to be changed or variables that are to be controlled. **Do not** confuse with control group.

dependent variable—The factor that changes as a result of altering the independent variable. Also, the change in events or results linked and controlled by another factor that has also been changed.

experiment—A planned investigation to determine the outcome that would arise from changing a variable or from changing "natural" conditions.

experimental group—A group of subjects to which independent or experimental variables are applied.

experimental variable—See *independent variable*.

graphs—Illustrated form of presenting raw, smooth, or analyzed data.

hypothesis—Statement of an idea that can be tested experimentally, based upon research. States what experimenter believes will happen as a result of the experiment.

independent variable—The item, quantity, or condition that is altered to observe what will happen; something that can be changed in an experiment without causing a change in other variables.

interpretation—One's personal viewpoint based on the data. This can be based on either qualitative or quantitative analysis, and may become a part of the project's conclusion.

materials—All items used in the course of the experiment.

measured variable—See *dependent variable*.

observation—What one sees in the course of the experiment. Observations are often incorporated into raw data.

procedures—Steps that must be followed to perform an experiment.

proposal—The planned procedure for experimentation.

qualitative analysis—Analysis made subjectively, without measurement.

quantitative analysis—Analysis made objectively, with measurement devices.

question (or problem)—That which forms the basis of the hypothesis and hence is the objective of the experiment.

raw data—Logs, tables, and graphs that represent data as it is collected in the course of the experiment.

research—The process of learning facts or prior theories on a subject by reviewing existing sources of information.

results—Graphs and tables that represent raw, smooth, and analyzed data.

scientific method—Manner of conducting an experiment, using valid subjects, variables and controls, and accurately recording results.

smooth data—Tables or graphs where all the averages, totals, or percentages are placed. These may combine the information from raw data tables and graphs.

tables—Written form of presenting raw, smooth, or analyzed data.

variable—A condition that is changed to test the hypothesis or a condition that changes as a result of testing the hypothesis.

Bibliography

Horowitz, Lois, *Knowing Where to Look*. Cincinnati, OH: Writers Digest Books, 1984

Science Service, *Abstracts—37th International Science and Engineering Fair*. Washington, D.C., 1986

Science Service, *Abstracts—36th International Science and Engineering Fair*. Washington, D.C., 1985

Science Service, *Abstracts—35th International Science and Engineering Fair*. Washington, D.C., 1984

Science Service, *Abstracts—34th International Science and Engineering Fair*. Washington, D.C., 1983

Seeber, Edward D., *A Style Manual for Students—Based on the MLA Style Sheet*. Bloomington, IN: Indiana University Press, 1964

Strunk, William and White, E. B., *The Elements of Style*. New York, NY: Macmillan Publishing Company, 1979

Stoltzfus, John C., and Young, Dr. Morris N., *The Complete Guide To Science Fair Competition*. New York, NY: Hawthorn Books, Inc., 1972

Van Deman, Barry A. and McDonald, Ed *Nuts & Bolts, A Matter of Fact Guide to Science Fair Projects*. Harwood Heights, IL: The Science Man Press, 1980

Index

Other Bestsellers From TAB

☐ **BUILD YOUR OWN WORKING ROBOT—THE SECOND GENERATION**—David L. Heiserman

Are you bored with ordinary, every-day types of electronic project? Ready for a real challenge? Meet Buster . . . an amazing, personal robot that you can build yourself. You'll get instructions on the two modes Buster operates under—his independent one in which he uses his blunder response to run freely without operator control . . . and dependent mode in which he is in the control of the person using the Operator's Control Panel. 140 pp., 37 illus.

Paper $12.95   **Hard $18.95**
Book No. 2781

☐ **62 HOME REMOTE CONTROL AND AUTOMATION PROJECTS**—Delton T. Horn

Here are 62 different remote control and automation units that anyone can build. All parts and components are readily obtainable and the cost is only a fraction of that charged for ready-built devices in electronics stores . . . some of the devices can't be purchased commercially at any price! Just think how you can impress your friends with your high-tech know-how while making your home safer and more convenient. 294 pp., 222 illus.

Paper $12.95   **Hard $18.95**
Book No. 2735

☐ **TIME GATE: HURTLING BACKWARD THROUGH HISTORY**—Charles R. Pellegrino

Taking a new approach to time travel, this totally fascinating history of life on Earth transports you backward from today's modern world through the very beginnings of man's existence. Interwoven with stories and anecdotes, and illustrated with exceptional drawings and photographs, this is history as it should always have been written! It will have you spellbound from first page to last! 288 pp., 142 illus. 7″ × 10″.

Paper $16.95   **Book No. 1863**

☐ **333 SCIENCE TRICKS AND EXPERIMENTS**—Robert J. Brown

Here is a delightful collection of experiments and "tricks" that demonstrate a variety of well-known, and not so well-known, scientific principles and illusions. Find tricks based on inertia, momentum, and sound projects based on biology, water surface tension, gravity and centrifugal force, heat, and light. Every experiment is easy to understand and construct . . . using ordinary household items. 208 pp., 189 illus.

Paper $9.95   **Hard $15.95**
Book No. 1825

☐ **THE COMPLETE PASSIVE SOLAR HOME BOOK**—Brad Schepp and Stephen M. Hastie

You'll get down-to-earth pointers on basic energy conservation . . . a clear picture of passive solar home design options—underground homes, superinsulated homes, double envelope houses, and manufactured housing. Plus, you'll get expert how-to's for choosing a passive solar design . . . how to deal with architects, designers, contractors . . . and more. 320 pp., 252 illus. 7″ × 10″.

Paper $16.95   **Book No. 1657**

☐ **CHIP TALK: PROJECTS IN SPEECH SYNTHESIS**—Dave Prochnow

Speech synthesis is one of today's most fascinating new technologies. And now it's one you can explore easily and inexpensively with the help of this excellent introduction to the principles of speech synthesis, complete with step-by-step instructions for constructing a variety of working synthesizers including both stand-alone and computer-interfaced units—at a fraction of the cost of comparable commercial units. 224 pp., 131 illus.

Paper $14.95   **Hard $24.95**
Book No. 2812

☐ **THE ELECTRONICS MANUAL TO INDUSTRIAL AUTOMATION**—G. Randy Slone

You'll go through a complete industrial training course on how automation works, its industrial electronic application, and how to repair and service automation equipment. Best of all, this book offers information that is understandable not only to those with engineering degrees, but to any industrial professional who must live with automation, maintain it, and effectively utilize it in the workplace. This guide will help you make intelligent decisions regarding the purchase and implementation of automation equipment. However, if you are a technically-oriented person, you'll find this guide extremely valuable in bringing you up to date with all the most modern types of automation systems. 352 pp., 178 illus.

Vinyl $17.95   **Book No. 2762**

☐ **PARTICLES IN NATURE: THE CHRONOLOGICAL DISCOVERY OF THE NEW PHYSICS**—John H. Mauldin

If you're interested in physics, science, astronomy, or natural history, you will find this presentation of the particle view of nature fascinating, informative, and entertaining. John Mauldin has done what few other science writers have been able to accomplish . . . he's reduced the complex concepts of particle physics to understandable terms and ideas. This enlightening guide makes particle physics seem less abstract—it shows significant spin-offs that have resulted from research done, and gives a glimpse of future research that promises to be of practical value to everyone. 304 pp., 169 illus. 16 Full-Color Pages, 14 Pages of Black & White Photos. Large Format (7″ × 10″).

Paper $16.95   **Hard $23.95**
Book No. 2616

☐ **333 *MORE* SCIENCE TRICKS AND EXPERIMENTS**—Robert J. Brown

Here's an ideal way to introduce youngsters of all ages to the wonders and complexities of science . . . a collection of tricks and experiments that can be accomplished with ordinary tools and materials. You can demonstrate that air is "elastic," perform hydrotropism, geotropism, plant force, phototropism, or psychological tricks . . . or perform any one of more than 300 fascinating experiments. 240 pp., 189 illus.

Paper $10.95   **Hard $15.95**
Book No. 1835

*Prices subject to change without notice.

Look for these and other TAB books at your local bookstore.

TAB BOOKS Inc.
P.O. Box 40
Blue Ridge Summit, PA 17214

Send for FREE TAB catalog describing over 1200 current titles in print.

Or Call For Immediate Service   1-800-233-1128

Pearisburg Public Library

Pearisburg, Virginia

1. Books may be kept two weeks and may be renewed once for the same period, except 7 day books and magazines.

2. A fine is charged for each day a book is not returned according to the above rule. No book will be issued to any person incurring such a fine until it has been paid.

3. All injuries to books beyond reasonable wear and all losses shall be made good to the satisfaction of the Librarian.

4. Each borrower is held responsible for all books charged on his card and for all fines accruing on the same.